W9-CLV-392

Motion, Sound, and Heat

From the ancient Greeks through the Age of Newton, the problems of motion, sound, and heat preoccupied the scientific imagination. These centuries gave birth to the basic concepts from which modern physics has evolved. In this first volume of his celebrated UNDERSTANDING PHYSICS, Isaac Asimov deals with this fascinating, momentous stage of scientific development with an authority and clarity that add further luster to an eminent reputation. Demanding the minimum of specialized knowledge from his audience, he has produced a work that is the perfect supplement to the student's formal textbook, as well as offering invaluable illumination to the general reader.

ABOUT THE AUTHOR: ISAAC ASIMOV is generally regarded as one of this country's leading writers of science and science fiction. He obtained his Ph.D. in chemistry from Columbia University and was Associate Professor of Biochemistry at Boston University School of Medicine. He is the author of over two hundred books, including *The Chemicals of Life, The Genetic Code, The Human Body, The Human Brain,* and *The Wellsprings of Life,* all available in Mentor editions.

MENTOR SCIENCE LIBRARY
Books by Isaac Asimov

☐ UNDERSTANDING PHYSICS: The Electron, Proton, and Neutron   (#ME1802—$2.25)

☐ UNDERSTANDING PHYSICS: Light, Magnetism, and Electricity   (#ME1792—$2.25)

☐ THE CHEMICALS OF LIFE   (#MW1606—$1.50)

☐ THE HUMAN BODY: Its Structure and Operation   (#MJ1774—$1.95)

☐ THE HUMAN BRAIN: Its Capacities and Functions   (#MJ1712—$1.95)

☐ WELLSPRINGS OF LIFE   (#MW1619—$1.50)

☐ WORDS OF SCIENCE   (#MJ1799—$1.95)

☐ THE GENETIC CODE   (#ME1821—$1.75)

Buy them at your local bookstore or use this convenient coupon for ordering.

THE NEW AMERICAN LIBRARY, INC.,
P.O. Box 999, Bergenfield, New Jersey 07621

Please send me the MENTOR BOOKS I have checked above. I am enclosing $_____(please add 50¢ to this order to cover postage and handling). Send check or money order—no cash or C.O.D.'s. Prices and numbers are subject to change without notice.

Name _____

Address _____

City_____ State_____ Zip Code_____
Allow 4-6 weeks for delivery.
This offer is subject to withdrawal without notice.

UNDERSTANDING PHYSICS

Volume I

Motion, Sound and Heat

ISAAC ASIMOV

A MENTOR BOOK
NEW AMERICAN LIBRARY
TIMES MIRROR
NEW YORK AND SCARBOROUGH, ONTARIO
THE NEW ENGLISH LIBRARY LIMITED, LONDON

NAL BOOKS ARE AVAILABLE AT QUANTITY DISCOUNTS
WHEN USED TO PROMOTE PRODUCTS OR SERVICES. FOR
INFORMATION PLEASE WRITE TO PREMIUM MARKETING
DIVISION, THE NEW AMERICAN LIBRARY, INC., 1633
BROADWAY, NEW YORK, NEW YORK 10019.

COPYRIGHT © 1966 BY ISAAC ASIMOV

All rights reserved. No portion of this work may be reproduced
without permission except for brief passages for the purpose of
review. For information address The New American Library, Inc.

*This is a reprint of a hardcover edition published by
Walker and Company. The hardcover edition was published
simultaneously in Canada by George J. McLeod, Limited,
Toronto.*

MENTOR TRADEMARK REG. U.S. PAT. OFF. AND FOREIGN COUNTRIES
REGISTERED TRADEMARK—MARCA REGISTRADA
HECHO EN CHICAGO, U.S.A.

SIGNET, SIGNET CLASSICS, MENTOR, PLUME, MERIDIAN AND NAL
Books are published *in the United States* by
The New American Library, Inc.,
1633 Broadway, New York, New York 10019,
in Canada by The New American Library of Canada Limited,
81 Mack Avenue, Scarborough, Ontario M1L 1M8,
in the United Kingdom by The New English Library Limited,
Barnard's Inn, Holborn, London, EC1N 2JR, England.

FIRST MENTOR PRINTING, APRIL, 1969

6   7   8   9   10   11   12   13   14

PRINTED IN THE UNITED STATES OF AMERICA

TABLE OF CONTENTS

15 Thermodynamics

1

The Search for Knowledge

From Philosophy to Physics

The scholars of ancient Greece were the first we know of to attempt a thoroughgoing investigation of the universe—a systematic gathering of knowledge through the activity of human reason alone. Those who attempted this rationalistic search for understanding, without calling in the aid of intuition, inspiration, revelation, or other nonrational sources of information, were the *philosophers* (from Greek words meaning "lovers of wisdom").*

Philosophy could turn within, seeking an understanding of human behavior, of ethics and morality, of motivations and responses. Or it might turn outside to an investigation of the universe beyond the intangible wall of the mind—an investigation, in short, of "nature."

Those philosophers who turned toward the second alternative were the *natural philosophers*, and for many centuries after the palmy days of Greece the study of the phenomena of nature continued to be called natural philosophy. The modern word that is used in its place—*science*, from a Latin word meaning "to

---

* Undoubtedly there were wise men, and even rationalists, before the Greeks, but they are not known to us by name. Furthermore, the pre-Greek rationalists labored in vain, for it was only the Greek culture that left behind it a rationalistic philosophy to serve as ancestor to modern science.

know"—did not come into popular use until well into the nineteenth century. Even today, the highest university degree given for achievement in the sciences is generally that of "Doctor of Philosophy."

The word "natural" is of Latin derivation, so the term "natural philosophy" stems half from Latin and half from Greek, a combination usually frowned upon by purists. The Greek word for "natural" is *physikos*, so one might more precisely speak of *physical philosophy* to describe what we now call science.

The term *physics*, therefore, is a brief form of physical philosophy or natural philosophy and, in its original meaning, included all of science.

However, as the field of science broadened and deepened, and as the information gathered grew more voluminous, natural philosophers had to specialize, taking one segment or another of scientific endeavor as their chosen field of work. The specialties received names of their own and were often subtracted from the once universal domain of physics.

Thus, the study of the abstract relationships of form and number became mathematics; the study of the position and movements of the heavenly bodies became astronomy; the study of the physical nature of the earth we live upon became geology; the study of the composition and interaction of substances became chemistry; the study of the structure, function, and interrelationships of living organisms became biology, and so on.

The term physics then came to be used to describe the study of those portions of nature that remained after the above-mentioned specialties were subtracted. For that reason the word has come to cover a rather heterogeneous field and is not as easy to define as it might be.

What has been left over includes such phenomena as motion, heat, light, sound, electricity, and magnetism. All these are forms of "energy" (a term about which I shall have considerably more to say later on), so that a study of physics may be said to include, primarily, a consideration of the interrelationships of energy and matter.

This definition can be interpreted either narrowly or broadly. If it is interpreted broadly enough, physics can be expanded to include a great deal of each of its companion sections of science. Indeed, the twentieth century has seen such a situation come about.

The differentiation of science into its specialties is, after all, an artificial and man-made state of affairs. While the level of knowledge was still low, the division was useful and seemed natural. It was possible for a man to study astronomy or biology without reference to chemistry or physics, or for that matter to study either chemistry or physics in isolation. With time and accumulated information, however, the borders of the specialties approached, met, and finally overlapped. The techniques of one science became meaningful and illuminating in another.

In the latter half of the nineteenth century, physical techniques made it possible to determine the chemical constitution and physical structure of stars, and the science of "astrophysics" was born. The study of the vibrations set up in the body of the earth by quakes gave rise to the study of "geophysics." The study of chemical reactions through physical techniques initiated and constantly broadened the field of "physical chemistry," and the latter in turn penetrated the study of biology to produce what we now call "molecular biology."

As for mathematics, that was peculiarly the tool of physicists (at first, much more so than that of chemists and biologists), and as the search into first principles became more subtle and basic, it became nearly impossible to differentiate between the "pure mathematician" and the "theoretical physicist."

In this book, however, I will discuss the field of physics in its narrow sense, avoiding consideration (as much as possible) of those areas that encroach on neighboring specialties.

The Greek View of Motion

Among the first phenomena considered by the curious Greeks was motion. One might initially suspect that motion is an attribute of life; after all, men and cats move freely but corpses and stones do not. A stone can be made to move, to be sure, but usually through the impulse given it by a living thing.

However, this initial notion does not stand up, for there are many examples of motion that do not involve life. Thus, the heavenly objects move across the sky and the wind blows as it wills. Of course, it might be suggested that heavenly bodies are pushed by angels and that wind is the breath of a storm-god, and indeed such explanations were common among most societies and through most centuries. The Greek philosophers, however, were committed to explanations that involved only that portion of the universe that could be deduced by human reason from phenomena apparent to human senses. That excluded angels and storm-gods.

Furthermore, there were pettier examples of motion. The

smoke of a fire drifted irregularly upward. A stone released in midair promptly moved downward, although no impulse in that direction was given it. Surely not even the most mystically-minded individual was ready to suppose that every wisp of smoke, every falling scrap of material, contained a little god or demon pushing it here and there.

The Greek notions on the matter were put into sophisticated form by the philosopher Aristotle (384–322 B.C.). He maintained that each of the various fundamental kinds of matter ("elements") had its own natural place in the universe. The element "earth," in which was included all the common solid materials about us, had as its natural place the center of the universe. All the earthy matter of the universe collected there and formed the world upon which we live. If every portion of the earthy material got as close to the center as it possibly could, the earth would have to take on the shape of a sphere (and this, indeed, was one of several lines of reasoning used by Aristotle to demonstrate that the earth was spherical and not flat).

The element "water" had its natural place about the rim of the sphere of "earth." The element "air" had its natural place about the rim of the sphere of "water," and the element "fire" had its natural place outside the sphere of "air."

While one can deduce almost any sort of scheme of the universe by reason alone, it is usually felt that such a scheme is not worth spending time on unless it corresponds to "reality"—to what our senses tell us about the universe. In this case, observation seems to back up the Aristotelian view. As far as the senses can tell, the earth is indeed at the center of the universe; oceans of water cover large portions of the earth; the air extends about land and sea; and in the airy heights there are even occasional evidences of a sphere of fire that makes itself visible during storms in the form of lightning.

The notion that every form of substance has its natural place in the universe is an example of an *assumption*. It is something accepted without proof, and it is incorrect to speak of an assumption as either true or false, since there is no way of proving it to be either. (If there were, it would no longer be an assumption.) It is better to consider assumptions as either useful or useless, depending on whether or not deductions made from them corresponded to reality.

If two different assumptions, or sets of assumptions, both lead to deductions that correspond to reality, then the one that explains more is the more useful.

On the other hand, it seems obvious that assumptions are the weak points in any argument, as they have to be accepted on faith in a philosophy of science that prides itself on its rationalism. Since we must start somewhere, we must have assumptions, but at least let us have as few assumptions as possible. Therefore, of two theories that explain equal areas of the universe, the one that begins with fewer assumptions is the more useful. Because William of Ockham (1300?–1349?), a medieval English philosopher, emphasized this point of view, the effort made to whittle away at unnecessary assumptions is referred to as making use of "Ockham's razor."

The assumption of "natural place" certainly seemed a useful one to the Greeks. Granted that such a natural place existed, it seemed only reasonable to suppose that whenever an object found itself out of its natural place, it would return to that natural place as soon as given the chance. A stone, held in the hand in midair, for instance, gives evidence of its "eagerness" to return to its natural place by the manner in which it presses downward. This, one might deduce, is why it has weight. If the supporting hand is removed, the stone promptly moves toward its natural place and falls downward. By the same reasoning, we can explain why tongues of fire shoot upward, why pebbles fall down through water, and why bubbles of air rise up through water.

One might even use the same line of argument to explain rainfall. When the heat of the sun vaporizes water ("turns it into air" a Greek might suppose), the vapors promptly rise in search of their natural place. Once those vapors are converted into liquid water again, the latter falls in droplets in search of their natural place.

From the assumption of "natural place," further deductions can be made. One object is known to be heavier than another. The heavier object pushes downward against the hand with a greater "eagerness" than the lighter object does. Surely, if each is released the heavier object will express its greater eagerness to return to its place by falling more rapidly than the lighter object. So Aristotle maintained, and indeed this too seemed to match observation, for light objects such as feathers, leaves, and snowflakes drifted down slowly, while rocks and bricks fell rapidly.

But can the theory withstand the test of difficulties deliberately raised? For instance, an object can be forced to move away from its natural place, as when a stone is thrown into the air. This is initially brought about by muscular impulse, but once the stone leaves the hand, the hand is no longer exerting an impulse upon it. Why then doesn't the stone at once resume its natural motion and fall to earth? Why does it continue to rise in the air?

Aristotle's explanation was that the impulse given the stone was transmitted to the air and that the air carried the stone along. As the impulse was transmitted from point to point in the air, however, it weakened, and the natural motion of the stone asserted itself more and more strongly. Upward movement slowed and eventually turned into a downward movement until finally the stone rested on the ground once more. Not all the force of an arm or a catapult could, in the long run, overcome the stone's natural motion. ("Whatever goes up must come down," we still say.)

It therefore follows that forced motion (away from the natural place) must inevitably give way to natural motion (toward the natural place) and that natural motion will eventually bring the object to its natural place. Once there, since it has no place else to go, it will stop moving. The state of *rest*, or lack of motion, is therefore the natural state.

This, too, seems to square with observation, for thrown objects come to the ground eventually and stop; rolling or sliding objects eventually come to a halt; and even living objects cannot move forever. If we climb a mountain we do so with an effort, and as the impulse within our muscles fades, we are forced to rest at intervals. Even the quietest motions are at some cost, and the impulse within every living thing eventually spends itself. The living organism dies and returns to the natural state of rest. ("All men are mortal.")

But what about the heavenly bodies? The situation with respect to them seems quite different from that with respect to objects on earth. For one thing, whereas the natural motion of objects here below is either upward or downward, the heavenly bodies neither approach nor recede but seem to move in circles about the earth.

Aristotle could only conclude that the heavens and the heavenly bodies were made of a substance that was neither earth, water, air, nor fire. It was a fifth "element," which he named "ether" (a Greek word meaning "blazing," the heavenly bodies being notable for the light they emitted).

The natural place of the fifth element was outside the sphere of fire. Why then, since they were in their natural place, did the heavenly bodies not remain at rest? Some scholars eventually answered that question by supposing the various heavenly bodies to be in the charge of angels who perpetually rolled them around the heavens, but Aristotle could not indulge in such easy explanations. Instead, he was forced into a new assumption to the effect that the laws governing the motion of heavenly bodies were different from those governing the motion of earthly bodies. Here the natural state was rest, but in the heavens the natural state was perpetual circular motion.

Flaws in Theory

I have gone into the Greek view of motion in considerable detail because it was a physical theory worked out by one of history's greatest minds. This theory seemed to explain so much that it was accepted by great scholars for two thousand years afterward; nevertheless it had to be replaced by other theories that differed from it at almost every point.

The Aristotelian view seemed logical and useful. Why then was it replaced? If it was "wrong," then why did so many people of intelligence believe it to be "right" for so long? And if they believed it to be "right" for so long, what eventually happened to convince them that it was "wrong"?

One method of casting doubt upon any theory (however respected and long established) is to show that two contradictory conclusions can be drawn from it.

For instance, a rock dropping through water falls more slowly than the same rock dropping through air. One might deduce that the thinner the substance through which the rock is falling the more rapidly it moves in its attempt to return to its natural place. If there were no substance at all in its path (a *vacuum*, from a Latin word meaning "empty"), then it would move with infinite speed. Actually, some scholars did make this point, and since they felt infinite speed to be an impossibility, they maintained that this line of argument proved that there could be no such thing as a vacuum. (A catch-phrase arose that is still current: "Nature abhors a vacuum.")

On the other hand, the Aristotelian view is that when a stone is thrown it is the impulse conducted by the air that makes it possible for the stone to move in the direction thrown. If the air were gone and a vacuum were present, there would be nothing to move the stone. Well then, would a stone in a vacuum move at infinite speed or not at all? It would seem we could argue the point either way.

Here is another possible contradiction. Suppose you have a one-pound weight and a two-pound weight and let them fall. The two-pound weight, being heavier, is more eager to reach its natural place and therefore falls more rapidly than the one-pound weight. Now place the two weights together in a tightly fitted sack and drop them. The two-pound weight, one might argue, would race downward but would be held back by the more leisurely fall of the one-pound weight. The overall rate of fall would therefore be an intermediate one, less than that of the two-pound weight falling alone and more than that of the one-pound weight falling alone.

On the other hand, you might argue, the two-pound weight and the one-pound weight together formed a single system weighing three pounds, which should fall more rapidly than the two-pound weight alone. Well then, does the combination fall more rapidly or less rapidly than the two-pound weight alone? It looks as though you could argue either way.

Such careful reasoning may point out weaknesses in a theory, but it rarely carries conviction, for the proponents of the theory can usually advance counter-arguments. For instance, one might say that in a vacuum natural motion becomes infinite in speed, while forced motion becomes impossible. And one might argue that the speed of fall of two connected weights depends on how tightly they are held together.

A second method of testing a theory, and one that has proved to be far more useful, is to draw a necessary conclusion from the theory and then check it against actual phenomena as rigorously as possible.

For instance, a two-pound object presses down upon the hand just twice as strongly as a one-pound object. Is it sufficient to say that the two-pound object falls more rapidly than the one-pound object? If the two-pound object displays just twice the eagerness to return to its natural place, should it not fall at just twice the rate? Should this not be tested? Why not measure the exact rate at which both objects fall and see if the two-pound object falls at just twice the rate of the one-pound object? If it doesn't, then surely the Greek theories of motion will have to be modified. If, on the other hand, the two-pound weight does fall just twice as rapidly, the Greek theories can be accepted with that much more assurance.

Yet such a deliberate test (or *experiment*) was not made by Aristotle or for two thousand years after him. There were two types of reasons for this. One was theoretical. The Greeks had had their greatest success in geometry, which deals with abstract concepts such as dimensionless points and straight lines without width. They achieved results of great simplicity and generality that they could not have obtained by measuring actual objects. There arose, therefore, the feeling that the real world was rather crude and ill-suited to helping work out abstract theories of the universe. To be sure, there were Greeks who experimented and drew important conclusions therefrom; for example, Archimedes (287?–212 B.C.) and Hero (first century A.D.). Nevertheless, the ancient and medieval view was definitely in favor of deduction from assumptions, rather than of testing by experimentation.

The second reason was a practical one. It is not as easy to experiment as one might suppose. It is not difficult to test the speed of a falling body in an age of stopwatches and electronic methods of measuring short intervals of time. Up to three centuries ago, however, there were no timepieces capable of measuring small intervals of time, and precious few good measuring instruments of any kind.

In relying on pure reason, the ancient philosophers were really making the best of what they had available to them, and in seeming to scorn experimentation they were making a virtue of necessity.*

The situation slowly began to change in the late Middle Ages. More and more scholars began to appreciate the value of experimentation as a method of testing theories, and here and there individuals began trying to work out experimental techniques.

The experimentalists remained pretty largely without influence, however, until the Italian scientist Galileo Galilei (1564–1642), came on the scene. He did not invent experimentation, but he made it spectacular and popular. His experiments with motion were so ingenious and conclusive that they not only began the destruction of Aristotelian physics but demonstrated the necessity, once and for all, of experimentalism in science. It is from Galileo (he is invariably known by his first name only) that the birth of "experimental science" or "modern science" is usually dated.

* And yet we can regret that the Greek philosophers did not conduct certain simple experiments that required no instruments. For instance, a sheet of thin papyrus falls slowly. The same sheet, crumpled into a small, tight ball, drops at a clearly greater speed. Since its weight hasn't changed as a result of the crumpling, why the change in the rate of fall? A question as simple as this might have been crucial in modifying Greek theories of motion in what we would now consider the proper direction.

CHAPTER 2

Falling Bodies

Inclined Planes

Galileo's chief difficulty was the matter of timekeeping. He had no clock worthy of the name, so he had to improvise methods. For instance, he used a container with a small hole at the bottom out of which water dripped into a pan at, presumably, a constant rate. The weight of water caught in this fashion between two events was a measure of the time that had elapsed.

This would certainly not do for bodies in "free fall"—that is, falling downward without interference. A free fall from any reasonable height is over too soon, and the amount of water caught during the time of fall is too small to make time measurements even approximately accurate.

Galileo, therefore, decided to use an inclined plane. A smooth ball will roll down a smooth groove on such an inclined plane at a manifestly lower speed than it would move if it were dropping freely. Furthermore, if the inclined plane is slanted less and less sharply to the horizontal, the ball rolls less and less rapidly; with the plane made precisely horizontal, the ball will not roll at all (at least, not from a standing start). By this method, one can slow the rate of fall to the point where even crude time-measuring devices can yield useful results.

One might raise the point as to whether motion down an

inclined plane can give results that can fairly be applied to free fall. It seems reasonable to suppose that it can. If something is true for every angle at which the inclined plane is pitched, it should be true for free fall as well, for free fall can be looked upon as a matter of rolling down an inclined plane that has been maximally tipped—that is, one that makes an angle of 90° with the horizontal.

For instance, it can be easily shown that relatively heavy balls of different weights would roll down a particular inclined plane at the same rate. This would hold true for any angle at which the inclined plane was tipped. If the plane were tipped more sharply, the balls would roll more rapidly, but all the balls would increase their rate of movement similarly; in the end all would cover the same distance in the same time. It is fair to conclude from that alone that freely falling bodies will fall through equal distances in equal times, regardless of their weight. In other words, a heavy body will *not* fall more rapidly than a light body, despite the Aristotelian view.

(There is a well-known story that Galileo proved this when he dropped two objects of different weight off the Leaning Tower of Pisa and they hit the ground in a simultaneous thump. Unfortunately, this is just a story. Historians are quite certain that Galileo never conducted such an experiment but that a Dutch scientist, Simon Stevinus (1548–1620), did something of the sort a few years before Galileo's experiments. In the cool world of science, however, careful and exhaustive experiments, such as those of Galileo with inclined planes, sometimes count for more than single, sensational demonstrations.)

Yet can we really dispose of the Aristotelian view so easily? The observed equal rate of travel on the part of balls rolling down an inclined plane cannot be disputed, but on the other hand neither is it possible to dispute the fact that a soap bubble falls far more slowly than a ping-pong ball of the same size, and that the ping-pong ball falls rather more slowly than a solid, wooden ball of the same size.

We have an explanation for this, however. Objects do not fall through nothing; they fall through air, and they must push the air aside, so to speak, in order to fall. We might take the viewpoint that to push the air aside consumes time. A heavy body pressing down hard pushes the light air to one side with no trouble and loses virtually no time. It doesn't matter whether the body is one pound or a hundred pounds. The one-pound weight experiences so little trouble in pushing the air to one side that the hundred-pound weight can scarcely improve on it. Both weights therefore fall through equal distances in equal times.* A distinctly light body such as a ping-pong ball would press down so softly that it would experience considerable trouble in pushing the air out of the way, and it would fall slowly. A soap bubble, for the same reason, would scarcely fall at all.

Can this use of air resistance as an explanation be considered valid? Or is it just something concocted to explain the failure of Galileo's generalization to hold in the real world? Fortunately, the matter can be checked. First, suppose that of two objects of equal weight one is spherical and compact while the other is wide and flat. The wide, flat object will make contact with air over a broader front and have to push more air out of the way in order to fall. It will therefore experience more air resistance than the spherical, compact one, and will fall more slowly, even though the two bodies are of equal weight. This turns out to be so, when tested. In fact, if a piece of paper is crumpled into a pellet, it falls more quickly because it suffers less air resistance. I referred to this experiment on page 9 as being one the Greeks might easily have performed, and from which they might have discovered that there must be something wrong with the Aristotelian view of motion.

An even more unmistakable test would be to get rid of air and allow bodies to fall through a vacuum. With no resistance to speak of, all bodies, however light or heavy they might be, ought to fall through equal distances in equal times. Galileo was convinced this would be so, but in his time there was no way of creating a vacuum to test the matter. In later years, when vacuums could be produced, the experiment of causing a feather and a lump of lead to fall together in a vacuum, and noting the fact that both covered an equal distance in an equal time, became commonplace. Air resistance is therefore real and not just a face-saving device.

Of course, this raises the question of whether one is justified, for the sake of enunciating a simple rule, in describing the universe in nonreal terms. Galileo's rule that all objects of whatever weight fell through equal distances in equal times could be expressed in very simple mathematical form. The rule is true, however, only in a perfect vacuum, which, as a matter of fact, does not exist. (Even the best vacuums we can create, even the vacuum of interstellar space, are not perfect.) On the other hand, Aristotle's view

* Actually, there is a small difference: This does not show up in falls of reasonable length, but would become visible if both weights were dropped from an airplane. In such case, the lighter weight would be held up a bit and lag behind a trifle.

that heavier objects fall more rapidly than light ones is true, at least to a certain extent, in the real world. However, it cannot be reduced to as simple a mathematical statement, for the rate of fall of particular bodies depends not only upon their weight but also upon their shapes.

One might suppose that reality must be held to at all costs. However, though that may be the most moral thing to do, it is not necessarily the most useful thing to do. The Greeks themselves chose the ideal over the real in their geometry and demonstrated very well that far more could be achieved by consideration of abstract line and form than by a study of the real lines and forms of the world; the greater understanding achieved through abstraction could be applied most usefully to the very reality that was ignored in the process of gaining knowledge.

Nearly four centuries of experience since Galileo's time has shown that it is frequently useful to depart from the real and to construct a "model" of the system being studied; some of the complications are stripped away, so a simple and generalized mathematical structure can be built up out of what is left. Once that is done, the complicating factors can be restored one by one, and the relationship suitably modified. To try to achieve the complexities of reality at one bound, without working through a simplified model first, is so difficult that it is virtually never attempted and, we can feel certain, would not succeed if it were attempted.

It is useless then to try to judge whether Galileo's views are "true" and Aristotle's "false" or vice versa. As far as rates of fall are concerned there are observations that back one view and other observations that back the other. What we can say, however, as strongly as possible, is that Galileo's views of motion turned out to explain many more observations in a far simpler manner than did Aristotle's views. The Galilean view was, therefore, far more useful. This was recognized not too long after Galileo's experiments were described, and Aristotelian physics collapsed.

Acceleration

If we were to measure the distance traversed by a body rolling down an inclined plane, we would find that the the body would cover greater and greater distances in successive equal time intervals.

Thus, a body might roll a distance of 2 feet in the first second. In the next second it would roll 6 feet, for a total distance of

8 feet. In the third second it would roll 10 feet, for a total distance of 18 feet. In the fourth second it would roll 14 feet, for a total distance of 32 feet.

Clearly the ball is rolling more and more rapidly with time.

This in itself represents no break with Aristotelian physics, for Aristotle's theories said nothing about the manner in which the velocity of a falling body changed with time. In fact, this increase in velocity might be squared with the Aristotelian view, for one might say that as a body approached its natural place its eagerness to get there heightened, so its velocity would naturally increase.

However, the importance of Galileo's technique was just this: he took up the matter of change of speed, not in a qualitative way but in a quantitative way. It is not enough to simply say, "Velocity increases with time." One must say, if possible, by just how much it increases and work out the precise interrelationship of velocity and time.

For instance, if a ball rolls 2 feet in one second, 8 feet in two seconds, 18 feet in three seconds, and 32 feet in four seconds, it would appear that there was a relationship between the total distance covered and the square of the time elapsed. Thus, 2 is equal to 2×1^2, 8 is equal to 2×2^2, 18 is equal to 2×3^2, and 32 is equal to 2×4^2. We can express this relationship by saying that the total distance traversed by a ball rolling down an inclined plane (or by an object in free fall) after starting from rest is directly proportional* to the square of the time elapsed.

Physics has adopted this emphasis on exact measurement that Galileo introduced, and other fields of science have done likewise wherever this has been possible. (The fact that chemists and biologists have not adopted the mathematical attitude as thoroughly as have physicists is no sign that chemists and biologists are less intelligent or less precise than physicists. Actually, this has come about because the systems studied by physicists are simpler than those studied by chemists and biologists and are more easily idealized to the point where they can be expressed in simple mathematical form.)

Now consider the ball rolling 2 feet in one second. Its average *velocity* (distance covered in unit time) during that one-

---

* When we say that *a* is "directly proportional" to *b*, we mean that as *b* increases, *a* increases as well. Sometimes, a relationship is such that as *b* increases, *a* decreases. (For instance, as the price of an object increases, the number of sales may decrease.) We then say that *a* is "inversely proportional" to *b*.

second interval is 2 feet divided by one second. It is easy to divide 2 by 1, but it is important to remember that we must divide the units as well, the "feet" by the "second." We can express this division of units in the usual fashion by means of a fraction. In other words, 2 feet divided by one second can be

expressed as $\frac{2 \text{ feet}}{1 \text{ second}}$, or 2 feet/second. This can be abbreviated

as 2 ft/sec and is usually read as "two feet per second." It is important not to let the use of "per" blind us to the fact that we are in effect dealing with a fraction. Its numerator and denominator are units rather than numbers, but the fractional quality remains nevertheless.

But to return to the rolling ball . . . In one second it covers 2 feet, for an average velocity of 2 ft/sec. In two seconds, it covers 8 feet, for an average velocity over the entire interval of 4 ft/sec. In three seconds it covers 18 feet, for an average velocity over the entire interval of 6 ft/sec. And you can see for yourself, the average velocity for the first four seconds is 8 ft/sec. The average velocity, all told, is in direct proportion to the time elapsed.

Here, however, we are dealing with average velocities. What is the velocity of a rolling ball at a particular moment? Consider the first second of time. During that second the ball has been rolling at an average velocity of 2 ft/sec. It began that first second of time at a slower velocity. In fact, since it started at rest, the velocity at the beginning (after 0 seconds, in other words) was 0 ft/sec. To get the average up to 2 ft/sec, the ball must reach correspondingly higher velocities in the second half of the time interval. If we assume that the velocity is rising smoothly with time, it follows that if the velocity at the beginning of the time interval was 2 ft/sec less than average, then at the end of the time interval (after one second), it should be 2 ft/sec more than average, or 4 ft/sec.

If we follow the same line of reasoning for the average velocities in the first two seconds, in the first three seconds, and so on, we come to the following conclusions: at 0 seconds, the velocity is 0 ft/sec; at one second, the velocity (at that moment) is 4 ft/sec; at two seconds, the velocity is 8 ft/sec; at three seconds, the velocity is 12 ft/sec; at four seconds, the velocity is 16 ft/sec, and so on.

Notice that after each second, the velocity has increased by exactly 4 ft/sec. Such a change in velocity with time is called an *acceleration* (from Latin words meaning "to add speed"). To determine the value of the acceleration, we must divide the gain in velocity during a particular time interval by that time interval. For instance at one second, the velocity was 4 ft/sec while at four seconds it was 16 ft/sec. Over a three-second interval the velocity increased by 12 ft/sec. The acceleration then is 12 ft/sec divided by three seconds. (Notice particularly that it is *not* 12 ft/sec divided by 3. Where units are involved, they *must* be included in any mathematical manipulation. Here you are dividing by three seconds and not by 3.)

In dividing 12 ft/sec by three seconds, we get an answer in which the units as well as the numbers are subjected to the division—in other words $4 \frac{\text{ft/sec}}{\text{sec}}$. This can be written 4 ft/sec/sec (and read four feet per second per second). Then again, in algebraic manipulations a/b divided by b is equal to a/b multiplied by $1/b$, and the final result is a/b^2. Treating unit-fractions in the same manner, 4 ft/sec/sec can be written 4 ft/sec^2 (and read four feet per second squared).

You can see that in the case just given, for whatever time interval you work out the acceleration, the answer is always the same: 4 ft/sec^2. For inclined planes tipped to a greater or lesser extent, the acceleration would be different, but it would remain constant for any one given inclined plane through all time intervals.

This makes it possible for us to express Galileo's discovery about falling bodies in simpler and neater fashion. To say that all bodies cover equal distances in equal times is true; however, it is not saying enough, for it doesn't tell us whether bodies fall at uniform velocities, at steadily increasing velocities, or at velocities that change erratically. Again, if we say that all bodies fall at equal velocities, we are not saying anything about how those velocities may change with time.

What we can say now is that all bodies, regardless of weight (neglecting air resistance), roll down inclined planes, or fall freely, at equal and constant accelerations. When this is true, it follows quite inevitably that two falling bodies cover the same distance in the same time, and that at any given moment they are falling with the same velocity (assuming both started falling at the same time). It also tells us that the velocity increases with time and at a constant rate.

Such relationships become more useful if we introduce mathematical symbols to express our meaning. In doing so, we introduce nothing essentially new. We would be saying in mathematical symbols exactly what we have been trying to say in words, but more briefly and more generally. Mathematics is a shorthand language in which each symbol has a precise and agreed-upon meaning. Once the language is learned, we find that it is only a form of English after all.

For instance, we have just been considering the case of an acceleration (from rest) of 4 ft/sec². This means that at the end of one second the velocity is 4 ft/sec, at the end of two seconds it is 8 ft/sec, at the end of the three seconds it is 12 ft/sec, and so on. In short, the velocity is equal to the acceleration multiplied by the time. If we let v stand as a symbol for "velocity" and t for "time," we can say that in this case v is equal to $4t$.

But the actual acceleration depends on the angle at which the inclined plane is tipped. If the plane is made steeper, the acceleration increases; if it is made less steep, the acceleration decreases. For any given plane, the acceleration is constant, but the particular value of the constant can vary greatly from plane to plane. Let us not, therefore, commit ourselves to a particular numerical value for acceleration, but let this acceleration be represented by a. We can then say:

$$v = at$$

(Equation 2–1)

It is important to remember that included in such equations in physics are units as well as numerals. Thus a, in Equation 2–1, does not represent a number merely, say 4, but a number and its units—4 ft/sec²—the unit being appropriate for acceleration. Again, t, for time, represents a number and its units—three seconds let us say. In evaluating at, then, we multiply 4 ft/sec² by three seconds, multiplying the units as well as the numerals. Treating the units as though they were fractions (in other words, as though we were to multiply a/b^2 by b) the product is 12 ft/sec. Thus, multiplying acceleration (a) by time (t) does indeed give us velocity (v), since the units we obtain, ft/sec, are appropriate to velocity.

In any equation in physics, the units on either side of the equals sign must balance after all necessary algebraic manipulation is concluded. If this balance is not obtained, the equation does not correspond to reality and cannot be correct. If the units of one symbol are not known, they can be determined by deciding just what kind of unit is needed to balance the equation. (This is sometimes called *dimensional analysis*.)

With that out of the way, let us consider a ball starting from rest and rolling down an inclined plane for *t* seconds. Since the ball starts at rest, its velocity at the beginning of the time interval is 0 ft/sec. According to Equation 2–1, at the end of the interval, at time *t*, its velocity *v* is *at* ft/sec. To get the average velocity, during this interval of smoothly increasing velocity, we take the sum of the original and final velocity $(0 + at)$ and divide by 2. The average velocity during the time interval is therefore $at/2$. The distance (*d*) traversed in that time must be the average velocity multiplied by the time, $at/2 \times t$. We therefore conclude that:

$$d = \frac{at^2}{2}$$

(Equation 2–2)

I will not attempt to check the dimensions for every equation presented, but let's do it for this one. The units of acceleration (*a*) are ft/sec² and the units of time (*t*) are sec (seconds). Therefore, the units of at^2 are ft/sec² × sec × sec, which works out to $\frac{\text{ft-sec}^2}{\text{sec}^2}$ or simply ft. Dividing at^2 by 2 does not alter the situation for in this case 2 is a "pure number"—that is, it lacks units. (Thus if you divide a foot-rule in two, each half has a length of 12 inches divided by 2, or 6 inches. The unit is not affected.) Thus the units of $at^2/2$ are ft (feet). an appropriate unit for distance (*d*).

Free Fall

As I said earlier, the value of the acceleration (*a*) of a ball rolling down an inclined plane varies according to the steepness of the plane. The steeper the plane, the greater the value of *a*.

Experimentation will show that for a given inclined plane the value of *a* is in direct proportion to the ratio of the height of the raised end of the plane to the length of the plane. If you represent the height of the raised end of the plane by *H*, and the length of the plane by *L*, you can express the previous sentence in mathematical symbols as $a \propto H/L$, where the symbol $\propto$ means "is directly proportional to."

In such a direct proportion the value of the expression on

one side changes in perfect correspondence with the value of the expression on the other. If H/L is doubled, a is doubled; if H/L is halved, a is halved; if H/L is multiplied by 2.529, a is multiplied by 2.529. This is what is meant by direct proportionality. But suppose that for a particular value of a, the value of H/L happens to be just a third as large. If the value of a is changed in any particular way, the value of H/L is changed in a precisely corresponding way, so it is still one third the value of a. In this particular case then, a is three times as large as H/L not for any one set of values but for all values.

This is a general rule. Whenever one factor, x, is directly proportional to another factor, y, we can always change the relationship into an equality by finding some appropriate constant value (usually called the *proportionality constant*) by which to multiply y. Ordinarily, we don't know the precise value of the proportionality constant to begin with, so it is signified by some symbol. This symbol is very often k (for "Konstant"—using the German spelling). Therefore, we can say that if $x \propto y$, then $x = ky$.

It is not absolutely necessary to use k as the symbol for the proportionality constant. Thus, the velocity of a ball rolling from rest is directly proportional to the time during which it has been rolling, and the distance it traverses is directly proportional to the square of that time; therefore, $v \propto t$ and $d \propto t^2$. In the first case, however, we have the special name "acceleration" for the proportionality constant, so we symbolize it by a; while in the second case, the relationship to acceleration is such that we symbolize the proportionality constant as $a/2$. Therefore $v = at$ and $d = at^2/2$.

In the case now under discussion, where the value of the acceleration (a) is directly proportional to H/L, it will prove convenient to symbolize the proportionality constant by the letter g. We can therefore say:

$$a = gH/L \qquad \text{(Equation 2-3)}$$

The quantities H and L are both measured in feet. In dividing H by L, feet are divided by feet and the unit cancels. The result is that the ratio H/L is a pure number and possesses no units. But the units of acceleration (a) are ft/sec². In order to keep the units in balance in Equation 2-3, it is therefore necessary that the units of g also be ft/sec², since H/L can contribute nothing in the way of units. We can conclude then that the proportionality constant in Equation 2-3 has the units of acceleration and therefore must represent an acceleration.

We can see what this means if we consider that the steeper we make a particular inclined plane, the greater the height of its raised end from the ground—that is, the greater the value of H. The length of the inclined plane (L) does not change, of course. Finally, when the plane is made perfectly vertical, the height of the raised end is equal to the full length of the plane, so that H equals L, and H/L equals 1.

A ball rolling down a perfectly vertical inclined plane is actually in free fall. Therefore, in free fall, H/L becomes 1, and Equation 2–3 becomes:

$$a = g \qquad \text{(Equation 2–4)}$$

This shows us that g is not only an acceleration, but is the particular acceleration undergone by a body in free fall. The tendency of a body to have weight and fall toward the earth is the result of a property called *gravity* (from the Latin word for "weighty"), and the symbol g is used because it is the abbreviation of "gravity."

If the actual acceleration of a body rolling down any particular inclined plane is measured, then the value of g can be obtained. Equation 2–3 can be rearranged to yield $g = aL/H$. For a particular inclined plane, the length (L) and height (H) of the raised end are easily measured, and with a known, g can be determined at once. Its value turns out to be equal to 32 ft/sec² (at least at sea level).

Now so far, for the sake of familiarity, I have made use of feet as a measure of distance. This is one of the common units of distance used in the United States and Great Britain, and we are accustomed to it. However, scientists all over the world use the *metric system* of measure, and we have gotten far enough into the subject, I think, to be able to join them in this.

The value of the metric system is that its various units possess simple and logical relationships among themselves. For instance, in the common system, 1 mile is equal to 1760 yards, 1 yard is equal to 3 feet, and 1 foot is equal to 12 inches. Converting one unit into another is always a chore.

In the metric system, the unit of distance is the "meter." Other units of distance are always obtained by multiplying the meter by 10 or a multiple of 10. Thanks to our system of writing numbers, this means that conversions of one unit to another within the metric system can be carried out by mere shifts of a decimal point.

Furthermore, standardized prefixes are used with set meaning. The prefix "deci-" always implies 1/10 of a standard unit, so a decimeter is 1/10 of a meter. The prefix "hecto-" always implies 100 times a standard unit, so a hectometer is 100 meters. And so it is for other prefixes as well.

The *meter* itself is 39.37 inches long. This makes it the equivalent, roughly, of 1.09 yards, or 3.28 feet. Two other metric units commonly used in physics are the *centimeter* and the *kilometer*. The prefix "centi-" implies 1/100 of a standard unit, so a centimeter is 1/100 of a meter. It is equivalent to 0.3937 inches, or approximately 2/5 of an inch. The prefix "kilo-" implies 1000 times the standard unit, so a kilometer is equal to 1000 meters or 100,000 centimeters. The kilometer is 39,370 inches long, which makes it just about 5/8 of a mile. The abbreviations ordinarily used for meter, centimeter, and kilometer are m, cm, and km, respectively.

Seconds, as a basic unit of time, are used in the metric system as well as in the common system. Therefore, if we want to express acceleration in metric units, we can use "meters per second per second" or m/sec^2 for the purpose. Since 3.28 feet equal 1 meter, we divide 32 ft/sec^2 by 3.28 and find that in metric units the value of g is 9.8 m/sec^2.

Once again, consider the importance of units. It is improper and incorrect to say that "the value of g is 32" or "the value of g is 9.8." The number by itself has no meaning in this connection. One must say either 32 ft/sec^2 or 9.8 m/sec^2. These last two values are absolutely equivalent. The numerical portions of the expression may be different, taken by themselves, but with the units added they are identical values. One is by no means "more true" or "more accurate" than the other; the expression in metric units is merely more useful.

We must know at all times which units are being used. In free fall, a is equal to g, so Equation 2–1 can be written $v = 32t$, if we are using common units; and $v = 9.8t$, if we are using metric units. In the shorthand of equations, the units are not included, so there is always the chance of confusion. If you try to use common units with the equation $v = 9.8t$, or metric units with the equation $v = 32t$, you will end up with results that do not correspond to reality. For that reason, the rules of procedure must be made perfectly plain. In this book, for instance, it will be taken for granted henceforward that the metric system will be used at all times, except where I specifically say otherwise.

Therefore, we can say that for bodies in free fall, from a starting position at rest:

$$v = 9.8t \qquad \text{(Equation 2–5)}$$

In the same way, for such a body, Equation 2–2 becomes $d = gt^2/2$ or:

$$d = 4.9t^2 \qquad \text{(Equation 2–6)}$$

At the end of one second, then, a falling body has dropped 4.9 m and is falling at a velocity of 9.8 m/sec. At the end of two seconds, it has fallen through a distance of 19.6 m and is falling at a velocity of 19.6 m/sec. At the end of the three seconds, it has fallen through a distance of 44.1 m and is falling at a velocity of 29.4 m/sec, and so on.*

* Since this book is not intended as a formal text, I am not presenting you with problems to be solved. I hope, nevertheless, that you have had enough experience with algebra to see that equations in physics not only present relationships in brief and convenient form, but also make it particularly convenient to solve problems—that is, to find the value of a particular symbol when the values of the other symbols in the equation are known or can be determined.

CHAPTER 3

The Laws of Motion

Inertia

Galileo's work on falling bodies was systematized a century later by the English scientist Isaac Newton (1642–1727), who was born, people are fond of pointing out, in the year of Galileo's death.

Newton's systematization appeared in his book *Philosophiae naturalis principia mathematica* (Mathematical Principles of Natural Philosophy) published in 1687. The book is usually referred to simply as the *Principia*.

Aristotle's picture of the physical universe had been lying shattered for nearly a hundred years, and it was Newton who now replaced it with a new one, more subtle and more useful. The foundations of the new picture of the universe consisted of three generalizations concerning motion that are usually referred to as Newton's Three Laws of Motion.*

* The important generalizations of science are brief descriptions of the behavior of the universe that are known to cover all observed cases. It is strongly felt that they will also cover all unobserved cases, here or anywhere, now or at any time. Such generalizations are sometimes called "laws of nature." This is actually a poor phrase because it seems to draw an analogy with man-made law, as something that is imposed and can be repealed, that can be violated at the cost of a penalty, and so on. All such analogies are misleading. It would be better therefore to speak of "Newton's generalizations concerning

His first law of motion may be given thus:

A body remains at rest or, if already in motion, remains in uniform motion with constant speed in a straight line, unless it is acted on by an unbalanced external force.

As you can see, this first law runs counter to the Aristotelian assumption of "natural place" with its corollary that the natural state of an object is to be at rest in its natural place.

The Newtonian view is that there is no natural place for any object. Wherever an object happens to be at rest† without any force acting upon it, it will remain at rest. Furthermore, if it happens to be in motion without any force acting upon it, it will remain in motion forever and show no tendency at all to come to rest. (I am not defining "force" just yet, but you undoubtedly already have a rough idea of what it means, and a proper definition will come eventually; see page 26.)

This tendency for motion (or for rest) to maintain itself steadily unless made to do otherwise by some interfering force can be viewed as a kind of "laziness," a kind of unwillingness to make a change. And indeed the first law of motion is referred to as the principle of *inertia*, from a Latin word meaning "idleness" or "laziness." (The habit of attributing human motivations or emotions to inanimate objects is called "personification." This is a bad habit in science, though quite a common one, and I indulged in it here only to explain the word "inertia.")

At first glance, the principle of inertia does not seem nearly as self-evident as the Aristotelian assumption of "natural place." We can see with our own eyes that moving objects do indeed tend to come to a halt even when, as nearly as we can see, there is nothing to stop them. Again, if a stone is released in midair it starts moving and continues moving at a faster and faster rate,

---

motion." However, everybody calls them the "laws" of motion, and if I did otherwise, I would merely seem eccentric. Nevertheless, by this footnote you are warned.

† In Aristotle's time the earth was considered a motionless body fixed at the center of the universe; the notion of "rest" therefore had a literal meaning. What we ordinarily consider "rest" nowadays is a state of being motionless with respect to the surface of the earth. But we know (and Newton did, too) that the earth itself is in motion about the sun and about its own axis. A body resting on the surface of the earth is therefore not really in a state of rest at all. In fact, the whole problem of what is really meant by "rest" and "motion" forced a new view of the universe in the form of what is called the "theory of relativity," advanced by Albert Einstein in 1905. In this book, however, we will run into no complications if we accept the fact that by "rest" and "motion" we really mean "rest with respect to the earth's surface" and "motion with respect to the earth's surface."

even though, as nearly as we can see, there is nothing to set it into motion.

If the principle of inertia is to hold good, we must be willing to admit the presence of subtle forces that do not make their existence very obvious.

For instance, a hockey puck given a sharp push along a level cement sidewalk will travel in a straight line, to be sure, but will do so at a quickly decreasing velocity and soon come to a halt. If the same puck is given the same sharp push along a smooth layer of ice, it will travel much farther, again in a straight line, but this time at only a slowly decreasing velocity. If we experiment sufficiently, it will quickly become clear that the rougher the surface along which the puck travels, the more quickly it will come to a halt.

It would seem that the tiny unevennesses of the rough surface catch at the tiny unevennesses of the hockey puck and slow it up. This catching of unevennesses against unevennesses is called *friction* (from a Latin word meaning "rub"), and the friction acts as a force that slows the puck's motion. The less the friction, the smaller the frictional force and the more slowly the puck's velocity is decreased. On a very smooth surface, such as that of ice, friction is so low that a puck would travel for great distances. If one could imagine a horizontal surface with no friction at all, then the hockey puck would travel in a straight line at constant velocity forever.

The Newtonian principle of inertia therefore holds exactly only in an imaginary ideal world in which no interfering forces exist: no friction, no air resistance.

Next consider a rock held in midair. It is at rest, but the instant we let go it begins to move. Clearly, then, there must be some force that makes it move, since the principle of inertia requires that in the absence of a force it remain at rest. Since the motion of the rock, if merely released, is always in the direction of the earth, the force must be exerted in that direction. Since the property that makes a rock fall had long been spoken of as "gravity," it was natural to call the force that brought about the motion *gravitational force* or the *force of gravity*.

It would therefore seem that the principle of inertia depends upon a circular argument. We begin by stating that a body will behave in a certain way unless a force is acting on it. Then, whenever it turns out that a body does not behave in that way, we invent a force to account for it.

Such circular argumentation would be bad indeed if we set

about trying to prove Newton's first law, but we do not do this. Newton's laws of motion represent assumptions and definitions and are not subject to proof. In particular, the notion of "inertia" is as much an assumption as Aristotle's notion of "natural place." There is this difference between them, however: The principle of inertia has proved extremely useful in the study of physics for nearly three centuries now and has involved physicists in no contradictions. For this reason (and not out of any considerations of "truth") physicists hold on to the laws of motion and will continue to do so.

To be sure, the new relativistic view of the universe advanced by Einstein makes it plain that in some respects Newton's laws of motion are only approximations. At very great velocities and over very great distances, the approximations depart from reality by a considerable amount. At ordinary velocities and distance, however, the approximations are extremely good.*

Forces and Vectors

The term *force* comes from the Latin word for "strength," and we know its common meaning when we speak of the "force of circumstance" or the "force of an argument" or "military force." In physics, however, force is defined by Newton's laws of motion. A force is that which can impose a change of velocity on a material body.

We are conscious of such forces, usually (but not always), as muscular effort. We are conscious, furthermore, that they can be exerted in definite directions. For instance, we can exert a force on an object at rest in such a way as to cause it to move away from us. Or we can exert a similar force in such a way as to cause it to move toward us. The forces are clearly exerted in different directions, and in common speech we give such forces two separate names. A force directed away from ourselves is a *push;* one directed toward ourselves is a *pull.* For this reason, a force is sometimes defined as "a push or a pull," but this is actually no definition at all, for it only tells us that a force is either one kind of force or another kind of force.

---

* It is sometimes said that Einstein's view of the universe "disproved" Newton's view. This is too simple a view. Actually, Einstein's view is more useful over a wider range of circumstances. Under ordinary circumstances, however, the Einsteinian view works out to be just about identical with the Newtonian view In this book, ordinary circumstances only will be involved, and it will not be necessary to introduce relativity

A quantity that has both size and direction, as force does, is a *vector quantity*, or simply a *vector*. One that has size only is a *scalar quantity*. For instance, distance is usually treated as a scalar quantity. An automobile can be said to have traveled a distance of 15 miles regardless of the direction in which it was traveling.

On the other hand, under certain conditions direction does make a difference when it is combined with the size of the distance. If town B is 15 miles north of town A, then it is not enough to direct a motorist to travel 15 miles to reach town B. The direction must be specified. If he travels 15 miles north, he will get there; if he travels 15 miles east (or any direction other than north), he will not. If we call a combination of size and direction of distance traveled *displacement*, then we can say that displacement is a vector.

The importance of differentiating between vectors and scalars is that the two are manipulated differently. For instance, in adding scalars it is sufficient to use the ordinary addition taught in grade school. If you travel 15 miles in one direction, then travel 15 miles in another direction, the total distance you travel is 15 plus 15, or 30 miles. Whatever the directions, the total mileage is 30.

If you travel 15 miles north, then another 15 miles north, the total displacement is, to be sure 30 miles north. Suppose, however, that you travel 15 miles north, then 15 miles east. What is your total displacement? How far, in other words, are you from your starting point? The total distance traveled is still 30 miles, but your final displacement is 21.2 miles northeast. If you travel 15 miles

Scalar and vector

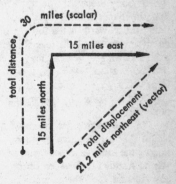

north and then 15 miles south, you have still traveled 30 miles altogether, but your total displacement is 0 miles, for you are back at your starting point.

So there is both ordinary addition, involving scalars, and *vector addition*, involving vectors. In ordinary addition $15 + 15$ is always 30; in vector addition, $15 + 15$ can be anything from 0 to 30, depending on circumstances.

Since force is a vector, two forces are added together according to the principles of vector addition. If one force is applied to a body in one direction and an exactly equal force is applied in the opposite direction, the sum of the two forces is zero; in such a case, even though forces are involved, a body subjected to them does not change its velocity. If it is at rest, it remains at rest. In fact, in every case where a body is at rest in the real world, we can feel certain that this does not mean there are no forces present to set it into motion. There are *always* forces present (the force of gravitation if nothing else). If there is rest, or unchanging velocity, that is because there is more than one force present and because the vector sum of all the forces involved is zero.

If the vector sum of all the forces involved is *not* zero, then there is an unbalanced force (mentioned in my definition of Newton's first law), or a net force. Whenever I speak of a force exerted on a body, it is to be understood that I mean a net force.

A particular force may have one of several effects on a moving body. The force of gravity, for instance, is directed downward toward the ground, and a falling body, moving in the direction of the gravitational pull, travels at a greater and greater velocity, undergoing an acceleration of 9.8 m/sec^2.

A body propelled upward, however, is moving in a direction opposite to that of the force of gravity. Consequently, it seems to be dragged backward by the force, moving more and more slowly. It finally comes to a halt, reverses its direction, and begins to fall. Such a slowing of velocity may be called "deceleration" or "negative acceleration." However, it would be convenient if a particular force was always considered to produce a particular acceleration. To avoid speaking of negative acceleration, we can instead speak of negative velocity.

In other words, let us consider velocity to be a vector. This means that a movement of 40 m/sec downward cannot be considered the same as a movement of 40 m/sec upward. The easiest way to distinguish between opposed quantities is to consider one positive and the other negative. Therefore, let us say that the downward motion is +40 m/sec and the upward one is —40 m/sec.

Since a downward force produces a downward acceleration* (acceleration being a vector, too), we can express the size of the acceleration due to gravity not as merely 9.8 m/sec, but as +9.8 m/sec.

If a body is moving at +40 m/sec (downward, in other words), the effect of acceleration is to increase the size of the figure. Adding two positive numbers by vector addition gives results similar to those of ordinary addition; therefore, after one second, the body is moving +49.8 m/sec, after another second, +59.6 m/sec, and so on. If, on the other hand, a body is moving at —40 m/sec (upward), the vector addition of a positive quantity resembles ordinary subtraction, as far as the figure itself is concerned. After one second, the body will be traveling —30.2 m/sec; after two seconds, —20.4 m/sec; and after four seconds —0.8 m/sec. Shortly after the four-second mark, the body will reach a velocity of 0 m/sec, and at that point it will come to a momentary halt. It will then begin to fall, and after five seconds its velocity will be +9.0 m/sec.

As can be seen, the acceleration produced by the force of gravity is the same whether the body is moving upward or downward, and yet there is something that is different in the two cases. The body covers more and more distance each second of its downward movement, but less and less distance each second of its upward movement. The amount of distance covered per unit time can be called the velocity or speed of the body.

In ordinary speech speed and velocity are synonymous, but not so in physics. Speed is a scalar quantity and does not involve direction. An object moving 16 m/sec north is traveling at the same speed as one moving at 16 m/sec east, but the two are traveling at different velocities. In fact, it is possible under certain circumstances to arrange a force so as to cause it to make a body move in circles. The speed, in that case, might not change at all, but the velocity (which includes direction) would be constantly changing.

Of the two terms, velocity is much more frequently used by physicists, for it is the broader and more convenient term. For

---

* We know from experience that if we push an object away from us, it moves away from us; if it is already moving, it moves away more rapidly. In the same way, to stop a moving body we always exert a force in the direction opposite to its motion. Experience tells us that the acceleration produced by a force is in the same direction as the force.

instance, we might define a force as "that which imposes a change in the speed of a body, or its direction of motion, or both." Or we might define it as "that which imposes a change in the velocity of a body," a briefer but as fully meaningful a phrase.

Since a change in velocity is an acceleration, we might also define a force as "that which imposes an acceleration on a body, the acceleration and force being in the same direction."

Mass

Newton's first law explains the concept of a force, but something is needed to allow us to measure the strength of a force. If we define a force as something that produces an acceleration, it would seem logical to measure the size of a force by the size of the acceleration it brings about. When we restrict ourselves to one particular body, say a basketball, this makes sense. If we push the basketball along the ground with a constant force, it moves more and more quickly, and after ten seconds it moves with a velocity, let us say, of 2 m/sec. Its acceleration is 2 m/sec divided by 10 seconds, or 0.2 m/sec². If you start from scratch and do not push quite as hard, at the end of ten seconds the basketball may be moving only 1 m/sec; it will therefore have undergone an acceleration of 0.1 m/sec². Since the acceleration is twice as great in the first case, it seems fair to suppose that the force was twice as great in the first case as in the second.

But if you were to apply the same forces to a solid cannonball instead of a basketball, the cannonball will not undergo anything like the previously noted accelerations. It might well take every scrap of force you can exert to get the cannonball to move at all.

Again, when a basketball is rolling along at 2 m/sec, you can stop it easily enough. The velocity change from 2 m/sec to 0 m/sec requires a force to bring it about, and you can feel yourself capable of exerting sufficient force to stop the basketball. Or you can kick the basketball in mid-motion and cause it to veer in direction. A cannonball moving at 2 m/sec, however, can only be stopped by great exertion, and if it is kicked in mid-motion it will change its direction by only a tiny amount.

A cannonball, in other words, behaves as though it possesses more inertia than a basketball and therefore requires correspondingly more force for the production of a given acceleration. Newton used the word *mass* to indicate the quantity of inertia possessed by a body, and his second law of motion states:

The acceleration produced by a particular force acting on a body is directly proportional to the magnitude of the force and inversely proportional to the mass of the body.

Now I have already explained that when x is said to be directly proportional to y, then $x=ky$ (see page 19). However, in saying that x is inversely proportional to another quantity, say z, we mean that as z increases x decreases by a corresponding amount and vice versa. Thus, if z is increased threefold, x is reduced to 1/3; if z is increased elevenfold x is reduced to 1/11, and so on. Mathematically, this notion of an inverse proportion is most simply expressed as $x \propto 1/z$, for then when z is 3, x is 1/3; when z is doubled to 6, x is halved to 1/6, and so on. We can change the proportionality to an equality by multiplying by a constant, so that if x is inversely proportional to z, $x = k/z$. If x is both directly proportional to y and inversely proportional to z, then $x = ky/z$.

With this in mind, let's have a represent the acceleration, f the magnitude of the force and m the mass of the body. We can then represent Newton's second law of motion as:

$$a = \frac{kf}{m} \qquad \text{(Equation 3–1)}$$

Let us next consider the units in which we will measure each quantity, turning to mass first, since we have not yet taken it into account in this book. You may think that if I say a cannonball is more massive than a basketball, I mean that it is heavier. Actually, I do not. "Massive" is not the same as "heavy," and "mass" is not the same as "weight," as I shall explain later in the book (see page 53). Nevertheless, there is a certain similarity between the two concepts and they are easily confused. In common experience, as bodies grow heavier they also grow more massive, and physicists have compounded the chance of confusion by using units of mass of a sort which nonphysicists usually think of as units of weight.

In the metric system, two common units for mass are the *gram* (gm) and the *kilogram* (kg). A gram is a small unit of mass. A quart of milk has a mass of about 975 grams, for example. The kilogram, as you might expect from the prefix, is equal to 1000 gm and represents, therefore, a trifle more than the mass of a quart of milk.

(In common units, mass is frequently presented in terms of "ounces" and "pounds," these units also being used for weight. In this book, however, I shall confine myself to the metric system as far as possible, and shall use common units, quarts, for example, only when they are needed for clarity.)

In measuring the magnitude of a force, two quantities must be considered: acceleration and mass. Using metric units, acceleration is most commonly measured as m/sec^2 or cm/sec^2, while mass may be measured in gm or kg. Conventionally, whenever distance is given in meters, the mass is given in kilograms, both being comparatively large units. On the other hand, whenever distance is given in the comparatively small centimeters, mass is given in the comparatively small grams. In either case, the unit of time is the second.

Consequently, the units of many physical quantities may be compounded of centimeters, grams, and seconds in various combinations; or of meters, kilograms, and seconds in various combinations. The former is referred to as the *cgs system*, the latter is the *mks system*. A generation or so ago, the cgs system was the more frequently used of the two, but now the mks system has gained in popularity. In this book, I will use both systems.

In the cgs system, a unit force is described as one that will produce an acceleration of 1 cm/sec^2 on a mass of 1 gm. A unit force is therefore 1 cm/sec^2 multiplied by 1 gm. (In multiplying the two algebraic quantities a and b, we can express the product simply as ab. We manipulate units as we would algebraic quantities, but to join words together directly would be confusing, so I will make use of a hyphen, which, after all, is commonly used to join words.) The product of 1 cm/sec^2 and 1 gm is therefore 1 gm-cm/sec^2—the magnitude of the unit force. The unit of force, gm-cm/sec^2, is frequently used by physicists, but since it is an unwieldy mouthful, it is more briefly expressed as the *dyne* (from a Greek word for "force").

Now let's solve Equation 3–1 for k. This works out to:

$$k = \frac{ma}{f} \qquad \text{(Equation 3–2)}$$

The value of k is the same for any consistent set of values of a, m and f, so we may as well take simple ones. Suppose we set m equal to 1 gm and a equal to 1 cm/sec^2. The amount of force that corresponds to such a mass and acceleration is, by our definition, 1 gm-cm/sec^2 (or 1 dyne).

Inserting these values into Equation 3–2, we find that:

$$k = \frac{1 \text{ cm/sec}^2 \times 1 \text{ gm}}{1 \text{ gm-cm/sec}^2} = \frac{1 \text{ gm-cm/sec}^2}{1 \text{ gm-cm/sec}^2} = 1$$

In this case, at least, k is a pure number.

Since k is equal to 1, we find that Equation 3-2 can be written as $ma/f = 1$, and, therefore:

$$f = ma \qquad \text{(Equation 3-3)}$$

provided we use the proper sets of units—that is, if we measure mass in gm, acceleration in cm/sec², and force in dynes.

In the mks system of measurement, acceleration is measured in m/sec² and mass in kg. The unit of force is then defined as that amount of force which will produce an acceleration of 1 meter per second per second when applied to 1 kilogram of mass. The unit force in this system is therefore 1 m/sec² multiplied by 1 kg, or 1 kg-m/sec². This unit of force is stated more briefly as 1 *newton*, in honor of Isaac Newton, of course. Equation 3-3 is still true, then, for a second combination of consistent units—where mass is measured in kg, acceleration in m/sec², and force in newtons.

From the fact that a kilogram is equal to 1000 grams and that a meter is equal to 100 centimeters, it follows that 1 kg-m/sec² is equal to (1000 gm) (100 cm)/sec², or 100,000 gm-cm/sec². To put it more compactly, 1 newton = 100,000 dynes.

Before leaving the second law of motion, let's consider the case of a body subject to no net force at all. In this case we can say that $f = 0$, so that Equation 3-3 becomes $ma = 0$. But any material body must have a mass greater than 0, so the only way in which ma can equal 0, is to have a itself equal 0.

In other words, if no net force acts on a body, it undergoes no acceleration and must therefore either be at rest or traveling at a constant velocity.

This last remark, however, is an expression of Newton's first law of motion. It follows, then, that the second law of motion includes the first law as a *special case*. If the second law is stated and accepted, there is no need for the first law. The value of the first law is largely psychological. The special case of $f = 0$, once accepted, frees the mind of the "common-sense" Aristotelian notion that it is the natural tendency of objects to come to rest. With the mind thus freed, the general case can then be considered.

Action and Reaction

A force, to exist, must be exerted by something and upon something. It is obvious that something cannot be pushed unless something else is pushing. It should also be obvious that something cannot push unless there is something else to be pushed. You cannot imagine pushing or pulling a vacuum.

A force, then, connects two bodies, and the question arises as to which body is pushing and which is being pushed. When a living organism is involved, we are used to thinking of the organism as originating the force. We think of ourselves as pushing a cannon-ball and of a horse as pulling a wagon, not of the cannonball as pushing us or the wagon as pulling the horse.

Where two inanimate objects are concerned, we cannot be so certain. A steel ball falling upon a marble floor is going to push against the floor when it strikes and therefore exert a force upon it. On the other hand, since the steel ball bounces, the floor must have exerted a force upon the ball. Whereas the force of the ball was exerted downward onto the floor, the force of the floor was exerted upward onto the ball.

In this and in many other similar cases there would seem to be two forces, equal in magnitude and opposite in direction. Newton made the generalization that this was always and necessarily true in all cases and expressed it in his third law of motion. This is often stated very briefly: "For every action, there is an equal and opposite reaction." It is for that reason that the third law is sometimes referred to as the "law of action and reaction."

Perhaps, however, this is not the best way of putting it. By speaking of action and reaction, we are still thinking of a living object exerting a force on some inanimate object that then responds automatically. One force (the "action") seems to be more important and to precede in time the other force (the "reaction").

But this is not so. The two forces are of exactly equal importance (from the standpoint of physics) and exist simultaneously. Either can be viewed as the "action" or the "reaction." It would be better, therefore, to state the law something like this:

Whenever one body exerts a force on a second body, the second body exerts a force on the first body. These forces are equal in magnitude and opposite in direction.

So phrased, the law can be called the "law of interaction."

The third law of motion can cause confusion. People tend to ask: "If every force involves an equal and opposite counterforce, why don't the two forces always cancel out by vector addition, leaving no net force at all?" (If that were so, then acceleration would be impossible and the second law would be meaningless.)

The answer is that two equal and opposite forces cancel out by vector addition when they are exerted on the same body. If a force were exerted on a particular rock and an equal and opposite force were also exerted on that same rock, there would be no net force; the rock, if at rest, would remain at rest no matter how large each force was. (The forces might be large enough to crush the rock to powder, but they wouldn't move the rock.)

The law of interaction, however, involves equal and opposite forces exerted on *two separate bodies*. Thus, if you exert a force on a rock, the equal and opposite force is exerted by the rock on you; the rock and you each receive a single unbalanced force. If you exert a force on a rock and let go of it at the same time, the rock, in response to this single force, is accelerated in the direction of that force—that is, away from you. The second force is exerted on you, and you in turn accelerate in the direction of that second force—that is, in the direction opposite to that in which the rock went flying. Ordinarily, you are standing on rough ground and the friction between your shoes and the ground (accentuated, perhaps, by muscular bracing) introduces new forces that keep you from moving. Your acceleration is therefore masked, so the true effect of the law of interaction may go unnoticed. However, if you were standing on very smooth, slippery ice and hurled a heavy rock eastward, you would go sliding westward.

In the same way, the gases formed by the burning fuel in a rocket engine expand and exert a force against the interior walls of the engine, while the walls of the engine exert an equal and opposite force against the gases. The gases are forced into an acceleration downward, so that the walls (and the attached rocket) are forced into an acceleration upward. Every rocket that rises into the air is evidence of the validity of Newton's third law of motion.

In these cases, the two objects involved are physically separate, or can be physically separated. One body can accelerate in one direction and the other in the opposite direction. But what of the case where the two bodies involved are bound together? What of a horse pulling a wagon? The wagon also pulls the horse in the opposite direction with an equal force. Yet horse and wagon do not accelerate in opposite directions. They are hitched together and both move in the same direction.

If the forces connecting wagon and horse were the only ones involved, there would indeed be no overall movement. A wagon and horse on very slippery ice would get nowhere, no matter how the horse might flounder. On ordinary ground, there are frictional effects. The horse exerts a force on the earth and the earth exerts a counterforce on the horse (and its attached wagon). Consequently, the horse moves forward and the earth moves backward.

The earth is so much more massive than the horse that its acceleration backward (remember that the acceleration produced by a force is inversely proportionate to the mass of the body being accelerated) is completely unmeasurable. We are aware only of the horse's motion, and so it seems to us that the horse is pulling the wagon. We find it hard to imagine that the wagon is also pulling at the horse.

CHAPTER 4

Gravitation

Combination of Forces

Newton had already turned his attention to an important and very profound question while still in his twenties. Did the laws of motion apply only to the earth and its environs, or did they apply to the heavenly bodies as well? The question first occurred to him on his mother's farm when he saw an apple fall from a tree* and began to wonder whether the moon was in the grip of the same force as the apple was.

It might seem at first thought that the moon could *not* be in the grip of the same force as the apple, since the apple fell to earth and the moon did not. Surely, if the same force applied to both, the same acceleration would affect both, and therefore both would fall. However, this is an oversimplification. What if the moon is indeed in the grip of the same force as the apple and therefore moving downward toward the earth; in addition, what if the moon also undergoes a second motion? What if it is the combination of two motions that keeps the moon circling the earth and never quite falling all the way?

This notion of an overall motion being made up of two or more component motions in different directions was by no means

* It did *not* hit him on the head, despite the hundreds of cartoons drawn by hundreds of cartoonists.

an easy concept for scientists to accept. When Nicholas Copernicus (1473–1543) first suggested that the earth moved about the sun (rather than vice versa), some of the most vehement objections were to the effect that if the earth rotated on its axis and (still worse) moved through space in a revolution about the sun, it would be impossible for anything movable to remain fixed to the earth's surface. Anyone who leaped up in the air would come down many yards away, since the earth beneath him would have moved while he was in the air. Those arguing in this manner felt that this point was so obvious as to be unanswerable.

Those who accepted the Copernican notion of the motion of the earth had to argue that it was indeed possible for an object to possess two motions at once: that a leaping man, while moving up and down, could also move with the turning earth and therefore come down on the same spot from which he leaped upward.

Galileo pointed out that an object dropped from the top of the mast of a moving ship fell to a point at the base of the mast. The ship did not move out from under the falling object and cause it to fall into the sea. The falling object, while moving downward, must also have participated in the ship's horizontal motion. Galileo did not actually try this, but he proposed it as what is today called a "thought experiment." Even though it was proposed only in thought, it was utterly convincing; ships had sailed the sea for thousands of years, and objects must have been dropped from mast-tops during all those years, yet no seaman had ever reported that the ship had moved out from under the falling object. (And of course, we can flip coins on board speeding jets these days and catch them as they come down without moving our hand. The coin participates in the motion of the jet even while also moving up and down.)

Why then did some scholars of the sixteenth and seventeenth centuries feel so sure that objects could not possess two different motions simultaneously? Apparently it was because they still possessed the Greek habit of reasoning from what seemed valid basic assumptions and did not always feel it necessary to check their conclusions against the real universe.

For instance, the scholars of the sixteenth century reasoned that a projectile fired from a cannon or a catapult was potentially subject to motions resulting from two causes—first the impulse given it by the cannon or catapult, and secondly, its "natural motion" toward the ground. Assuming, to begin with, that an object could not possess two motions simultaneously, it would seem necessary that one motion be completed before the second began. In other words, it was felt that the cannonball would travel in a straight line in whatever direction the cannon pointed, until the impulse of the gunpowder explosion was used up; it would then at once fall downward in a straight line.

Galileo maintained something quite different. To be sure, the projectile traveled onward in the direction in which it left the cannon. What's more, it did so at constant velocity, for the force of the gunpowder explosion was exerted once and no more. (Without a continuous force there would be no continuous acceleration, Newton later explained.) *In addition,* however, the cannonball began dropping as soon as it left the cannon's mouth, in accordance with the laws of falling bodies whereby its velocity downward increased with a constant acceleration (thanks to the continuous presence of a constant force of gravity). It was easy to show by geometric methods that an object that moved in one direction at a constant speed, and in another at a speed that increased in direct proportion with time, would follow the path of a curve called a "parabola." Galileo also showed that a cannonball would have the greatest range if the cannon were pointed upward at an angle of 45° to the ground.

A cannon pointed at a certain angle would deliver a cannonball to one place if the early views of the cannonball's motions were correct, and to quite another place if Galileo's views were correct. It was not difficult to show that it was Galileo who was correct.

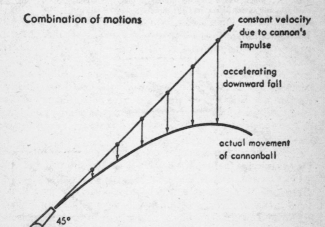

Combination of motions

constant velocity due to cannon's impulse

accelerating downward fall

actual movement of cannonball

45°

Indeed, the gunners of the time may not have dabbled much in theory, but they had long aimed their weapons in such a way as to take advantage of a parabolic motion of the cannonball.* In short, the possibility of a body's possessing two or more motions at once was never questioned after the time of Galileo.

How can separate motions be added together and a resultant motion obtained? This can be done by vector addition, according to a method most easily presented in geometric form. Consider two motions in separate directions, the two directions at an angle α to each other. (The symbol α is the Greek letter "alpha." Greek letters are often used in physics as symbols, in order to ease the overload on ordinary letters of the alphabet.) The two motions can then be represented by two arrows set at angle α, the two arrows having lengths in proportion to the two velocities. (If the velocity of one is twice that of the other, then its corresponding arrow is twice as long.) If the two arrows are made the sides of a parallelogram, the resultant motion built up out of the two component motions is

* The correct aim, especially nowadays when shells are hurled for miles, requires more than the idealized parabola of Galileo. Many factors of the real world—as, for instance, the curvature of the earth's surface, the manner in which its speed of rotation varies with latitude, the amount of air resistance (which varies with height and temperature), the strength and direction of the wind, the motion of the object aimed at and the object carrying the cannon (if both are ships, for instance), affect the situation. All these effects merely serve to modify the parabola, however, and do not affect the basic worth of Galileo's argument, which serves only to present a greatly simplified but nevertheless vastly useful model of the real situation.

Parallelogram of forces

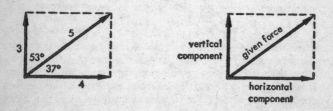

represented by the diagonal of the parallelogram, the one that lies in a direction intermediate between those of the two components.

Given the values of the two velocities and the angle between them, it is possible to calculate the size and direction of the resultant velocity even without the geometric construction, although the latter is always useful to lend visual aid. For instance, if one velocity is 3 m/sec in one direction, and the other is 4 m/sec in a direction at right angles to the first, then the resultant velocity is 5 m/sec in a direction that makes an angle of just under 37° with the larger component and just over 53° with the smaller.

In the same way, a particular velocity can be separated into two component velocities. The particular velocity is made the diagonal of a parallelogram, and the adjacent sides of the parallelogram represent the component velocities. This can be done in an infinite number of ways, since the line representing a velocity or force can be made the diagonal of an infinite number of parallelograms. As a matter of convenience, however, a velocity is divided into components that are at right angles to each other. The parallelogram is then a rectangle.

This device of using a parallelogram can be employed for the combination or resolution of any vector quantity. It is very frequently used for forces, as a matter of fact, so one usually speaks of this device as involving a *parallelogram of force*.

The Motion of the Moon

Now let us return to the moon. It travels about the earth in an elliptical orbit. The ellipse it describes in its revolution about the earth is not very far removed from a circle, however. The moon travels in this orbit with a speed that is almost constant.

Although the moon's speed is approximately constant, its velocity certainly is not. Since it travels in a curved path, its direction of motion changes at every instant and, therefore, so does its velocity. To say that the moon is continually changing its velocity is, of course, to say that it is subject to a continuing acceleration.

If the moon is viewed as traveling at a constant speed along a uniformly circular path (which is at least approximately true), it can be considered to be changing its direction of motion by precisely the same amount in each successive unit of time. It is therefore undergoing a constant acceleration and must be subject to a constant force, according to Newton's second law of motion. Since the shift in the direction of motion is always toward the earth, the acceleration, and therefore the force, must be directed toward the earth.

Certainly, if there is a force attracting the moon to the earth, it might well be the same as the force attracting the apple to the earth. However, if that were so, and the moon were undergoing a constant acceleration toward the earth under the pull of a constant force, why does it not fall to the earth as an apple would?

We can see why if we break the moon's motion into two component motions at mutual right angles. One of the components can be drawn as an arrow pointing toward the earth, along a radius

Motion of the moon

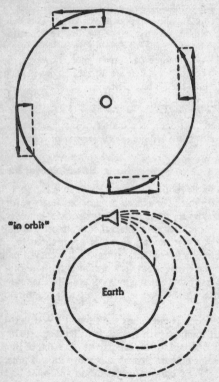

"in orbit"

Earth

of the moon's circular orbit. This represents the motion in response to the force attracting the moon to the earth. The other component is drawn at right angles to the first and is therefore tangent to the circle of the moon's orbit. This tangential motion represents that which the moon would experience if there were no force attracting it to the earth. The actual motion lies between the two, and the tangential component carries the moon to one side just far enough in a unit of time to make up for the motion toward the earth in that same unit of time. The moon, in other words, is always falling toward the earth, but it also "sidesteps."

In a sense, this "sidestep" means that the earth's surface curves away from the moon just as fast as the moon approaches by falling, and the distance between earth and moon remains the same. This can be made plain if one supposes a projectile fired horizontally from a mountaintop on earth with greater and greater velocity. The greater the velocity, the farther the projectile travels before striking the ground. The farther it travels, the more the surface of the spherical earth curves away from it, thus adding to the distance the projectile covers. Finally, if the projectile is shot forward with sufficient velocity, its rate of fall just matches the rate at which the earth's surface curves away, and the projectile "remains in orbit." It is in this fashion that satellites are placed in orbit, and it is in this fashion that the moon remains in orbit.

In considering the moon's motion, therefore, we need only consider that component which is directed toward the earth, and we can ask ourselves whether that component is the result of the same force that attracts the apple. Let's first concentrate on the apple and see how to interpret the force between it and the earth in the light of the laws of motion.

In the first place, all apples fall with the same acceleration regardless of how massive they are. But if one apple has twice the mass of a second apple, yet falls at the same acceleration, the first apple must be subjected to twice the force, according to the second law of motion. The force attracting the apple to the earth (often spoken of as the *weight* of the apple) must be proportional to the mass of the apple.

But according to the third law of motion, whenever one body exerts a force on a second, the second is exerting an equal and opposite force on the first. This means that if the earth attracts the apple with a certain downward force, the apple attracts the earth with an equal upward force.

That seems odd. How can a tiny apple exert a force equal

to that exerted by the tremendous earth? If it did, one would expect the apple to attract other objects as the earth does, and the apple most certainly does not. The logical way to explain this is to suppose that the attractive force between apple and earth depends not only on the mass of the apple but on the mass of the earth as well. It cannot depend on the sum of the masses, for when the mass of the apple is doubled, the sum of the mass of the apple and the earth remains just about the same as before, and yet the force of attraction doubles. Instead it must depend upon the product of the masses.

If we multiply the masses, the small mass has just as much effect on the final product as the large one. Thus, the minute quantity a multiplied by the tremendous quantity b yields the product ab. If a is now doubled, it becomes equal to $2a$. If that is multiplied by b, the product is $2ab$ Thus doubling one of two factors in a multiplication, however small that factor may be, doubles the product And doubling the mass of the apple doubles the size of the force between the apple and the earth

Furthermore, the apple does not measurably attract any other object of ordinary size because the product of the masses of two ordinary objects is an infinitesimal fraction of the product of the mass of either object and that of the vast earth. The attractive force between two objects of ordinary size is correspondingly smaller, and while the force does exist, it is far too small to be noticed in the ordinary course of events.

Since the earth attracts all material objects to itself (even the gaseous atmosphere is held firmly to the planet through gravitational force) it would seem that the force is produced by mass in whatever form the latter occurs. In that case, the earth need not be involved. Any two masses ought to interact gravitationally, and if we notice the force only when the earth is involved, that is only because the earth itself is the only body in our neighborhood massive enough to produce a gravitational force sufficient to obtrude itself on our notice.

Such is the essence of Newton's contribution. He did not discover the law of gravity merely in the sense that all earthly objects are attracted to the earth. (This limited concept is at least as old as Aristotle and the word "gravity" was used in that sense for many centuries before Newton.) What Newton pointed out was that *all* masses attracted *all other* masses, so that the earth's attraction was not unique. Because Newton maintained that there was a gravitational attraction between any two material bodies in the universe, his generalization is called the *law of universal gravitation*. The adjective "universal" is the most important word of the phrase.*

If this were all, we would now be able to decide the size of that component of the moon's motion that is directed toward the earth. All bodies on earth fall with the same acceleration, and therefore it might be decided that the moon, if it were in the grip of the same force, ought to do the same. In one second it ought to fall some 4.9 meters toward the earth. Actually, the earthward component of the moon's motion is much smaller than that.

To account for this, one might suppose that the earth's gravitational force weakens with distance, and certainly this seems a reasonable supposition. It is common experience that many things weaken with distance. Such is the case with light and sound, to name two common phenomena with which man has always been familiar.

And yet is such weakening supported by experimental evidence? At first blush it might seem that it wasn't. A stone dropped from a height of 100 meters falls with an acceleration of 9.8 m/sec², and one dropped from a height of 200 meters falls with the same acceleration. If the gravitational force decreased with distance from the earth, ought not the fall from a greater height involve a smaller acceleration? In fact, ought not the acceleration increase steadily as the stone approached the earth, instead of remaining constant, as it does?

But Newton's view was that *all* bodies attracted *all other* bodies. A falling rock is attracted not only by the portion of the earth making up the surface immediately under it, but also by the portions deep underneath, all the way to the center and beyond— to the antipodes, 12,740 kilometers (8000 miles) distant. It should also be attracted by portions at all distances to the north, east, south, west and points in between.

It would seem reasonable that for a body like the earth, which has nearly the symmetrical shape of a sphere, we could simplify matters. The pull from the north would balance the pull from the south; the pull from the west would balance the pull from the east; the distant pull of the antipodes would balance the nearby pull of the surface directly beneath. In consequence, we might suppose that the net effect is that the overall pull of the earth would be concentrated exactly at its center.†

* It is possible to contrive an exception. If the earth were hollow, there would be no net gravitational force within the hollow. A body within the hollow would not be attracted by the earth. However, this is a highly artificial exception that has no practical significance, for any body that is large enough to have an important gravitational field is too large to support a hollow structure.

† It is so easy to say "it would seem reasonable" and end with the happy

The radius of the earth is about 6370 kilometers (3960 miles). An object falling from a height of 100 meters (0.1 kilometers) begins its fall, therefore, from a point 6370.1 kilometers from the earth's center, while one falling from a height of 200 meters begins its fall from a point 6370.2 kilometers from the center. The difference is so insignificant that the gravitational attraction can be considered constant over that small distance. (Actually, modern instruments can measure the difference in the strength of the gravitational field over even such small distances with considerable accuracy.)

However, the moon's distance from the earth (center to center) is, on the average, 384,500 kilometers (239,000 miles). This is 60.3 times as far from the earth's center as is an object on the earth's surface. With a sixtyfold increase in distance, gravitational force might indeed decrease considerably.

But how much is "considerably"?

The earth attracts bodies on every portion of its surface; therefore, the gravitational force may be considered as radiating outward from the earth in all directions. If the force does this, it can be viewed as occupying the surface of a sphere that is inflating to a larger and larger size as it recedes from the earth. If a fixed amount of gravitational force is stretched out over the surface of such a growing sphere, then the intensity of the force at a given spot on the surface ought to decrease as the total surface area grows larger.

From solid geometry it is known that the surface area of a sphere is directly proportional to the square of its radius. If one sphere has three times the radius of another, it has nine times the surface area. As distance between two bodies increases, then the gravitational force between them ought to be inversely proportional to the square of that distance. (This relationship is familiarly known as the *inverse-square law*. Not only gravitation but also such phenomena as the intensity of light, the intensity of magnetic attraction, and the intensity of electrostatic attraction weaken as the square of the distance.)

In comparing the motion of the moon to the motion of an

conclusion that the gravitational force of the earth seems to originate at its center. Yet it took the transcendent genius of Newton eighteen years to convince himself of this fact, and he had to invent that branch of mathematics now known as the calculus, before he could prove it to his own satisfaction and that of others. Throughout this book, I say "it would seem reasonable" and "it is clear" and "it is easy to see" when I am reaching conclusions that in actual fact were attained only through great ingenuity and hard labor. In doing so, my conscience hurts—but in an introductory book I have no alternative.

apple on the earth's surface, we must remember that the moon is 60.3 times as far from the earth's center as the apple is and that the gravitational force on the moon is weaker by a factor of 60.3 times 60.3, or 3636. Whereas an apple falls 4.9 meters in the first second of fall, the moon should fall 1/3636 that distance, or 0.0013 meters, in a second of fall. (A thousandth of a meter is a *millimeter*, so that 0.0013 meters is equal to 1.3 millimeters.)

Indeed, astronomical measurements show that the moon in its course about the earth does indeed deviate from a straight line course by about 1.3 millimeters in one second. This alone would have been sufficient to make it strongly probable that the same force that held the apple held the moon. However, Newton went on to show how gravitational force on a universal scale would account for the fact that the orbit of the moon about the earth is an ellipse with the earth at one focus; that the planets revolved about the sun in a similar elliptical manner; that the tides took place as they did; that the precession of equinoxes took place, and so on. The one simple and straightforward generalization explained so much that it had to be triumphantly accepted by the scientific community.

A century after Newton's death, the German-English astronomer William Herschel (1738–1822) discovered instances of far distant stars that revolved about each other in strict accordance with Newton's law of universal gravitation, which thus seemed universal indeed. Unseen planets were eventually discovered through the tiny gravitational effects produced by their otherwise unsuspected presence. It is no wonder that Newton's working out of the law of universal gravitation is often considered as the greatest single discovery in the history of science.[*]

The Gravitational Constant

Newton succeeded in establishing the generalization that any two bodies in the universe attract each other with a force (F) that is directly proportional to the product of the masses (m

[*] Nevertheless, Newton's generalization concerning gravity is only an approximation and is not absolutely correct. Already in the mid-nineteenth century, it was discovered that the planet Mercury had one small component of its motion that could not be explained by Newton's law. It remained unexplained until Albert Einstein advanced his "General Theory of Relativity" in 1915. This theory, more advanced, powerful and controversial than the Special Theory of 1905, involved a broader view of the universe than was implicit in Newton's laws. In all ordinary cases, the two views were just about equivalent. At certain extremes, however, the two views diverge, and when such extremes are tested it appears that Einstein's view, rather than Newton's, carries the day.

and m') of the bodies, and inversely proportional to the square of the distance (d) between them. To convert the proportionality to an equality, it is necessary, of course, to introduce a constant. The one introduced in this case is usually referred to as the *gravitational constant* and is symbolized as G. Newton's law of universal gravitation can then be expressed as:

$$F = \frac{Gmm'}{d^2} \qquad \text{(Equation 4–1)}$$

One problem left unsolved by Newton was the value of G.

To see why it was left unsolved, let's consider the famous case of Newton's falling apple and try to substitute values in Equation 4–1, in the mks system of units. We know the value of the distance from the apple to the earth's center and can set d equal to 6,370,-000 meters. There are ways of measuring the mass of the apple and m can be set at, let us say, 0.1 kilograms. As for the strength of the gravitational force (F) between the apple and the earth, it is equal (see Equation 3–3) to the product of the mass of the apple and the acceleration to which it is subjected by the action of gravity, according to Newton's second law of motion. The value of F, then, is 0.1 kg times 9.8 m/sec², or 0.98 kg-m/sec².

This leaves us with two items still undetermined: G, the gravitational constant, and m', the mass of the earth. If we knew either one, we could calculate the other at once, but Newton knew neither, nor did anyone else in his time.

(You might wonder whether we could not eliminate the constant in Equation 4–1, as we did the constant in Equation 3–3. That, however, was done by a proper choice of units. We could do so here by inventing a unit called "earth-unit," for example, and saying that the earth had a mass of 1 earth-unit. We could further invent similar arbitrary units for the mass of the apple and the distance of the apple from the center of the earth. Such tricks would be of limited value, however. It is unsatisfying to be told that the earth has a mass of 1 earth-unit, and that is all we would find out in this way. What we want is the mass of the earth in terms of familiar objects—that is, in the units of the mks system. And for that we must know the value of G in the mks system.)

The law of universal gravitation implies that the value of G is the same under all conditions. Therefore, if we could measure the gravitational force between two bodies of known mass, separated by a known distance, we could at once determine G and from that the mass of the earth.

Unfortunately, the force of gravity is just about the weakest known force in nature. It takes a body with the enormous size of the earth to produce enough gravitational force to bring about an acceleration of 9.8 m/sec^2. The puny forces that can be produced by a few pounds of muscle can and do counter all that gravitational force whenever we chin ourselves, do pushups, jump upward, or climb a mountain.

For bodies that are large, though less massive than the earth, the decline in gravitational force has drastic effects. The earth maintains a firm grip on its atmosphere through the force of its gravity, but the planet Mars, which has only 1/10 the earth's mass, can hold only a thin atmosphere. The moon has an enormous mass by ordinary standards; nevertheless it is only 1/81 as massive as the earth and has a gravitational force too weak to hold any atmosphere at all.

Where bodies of ordinary size are concerned, the gravitational forces produced are completely insignificant. The mass in a mountain exerts a ·gravitational attraction on you, but you are aware of no difficulty in stepping away from such a mountain.

The problem, therefore, is how to measure so weak a force as that of gravity. We might speculate on possible ways of measuring the gravitational forces between two neighboring mountains, but the masses of individual mountains are not much easier to determine than the mass of the earth. Furthermore, the mountains are of irregular shape and the gravitational force is concentrated at some "center," a position that would be difficult to determine.

We would therefore have to measure the gravitational forces originating in symmetrical bodies small enough to be handled easily in the laboratory, and the measurement of the tiny gravitational forces to which such bodies would give rise might well be considered too difficult to lie within the realm of the possible.

The beginning of a solution to the problem came about, however, in the time of Newton himself, thanks to the work of the English scientist Robert Hooke (1635–1703).

As a preliminary to explaining Hooke's work, let us keep in mind that when forces are applied to a body, that body will change its shape as a result. If a plank of wood is suspended across two supports, and someone sits down on the center, that plank will bend under the load. If a rubber band is pulled at both ends in opposite directions, it stretches. If a sponge is clenched in a fist, it compresses, and if rotated at each end in opposite directions, it will twist. If pushed to the right at one end and to the left at the other, without being allowed to rotate, it will shear.

All these types of deforming forces can be referred to as *stresses*. The deformation undergone by the body under stress is a *strain*.

When an object undergoes deformation as the result of a stress, the original shape may be restored when the stress is removed. The wooden plank, after you stand up, unbends; the rubber band, after the pull is released, contracts to its normal size; the sponge once released from the compressive, twisting, or shearing force, springs back. Again, a steel ball flattens upon striking the ground, a baseball on striking the bat, and a golf ball on striking the club. When the deforming force is gone all are spheres once more. This tendency to return to the original shape after deformation under stress is called *elasticity*.

There is a limit to the elasticity of any substance, a point beyond which stress will produce a permanent deformation. For a substance such as wax, this point is easily reached and even light stresses will cause a lump of wax to change its shape permanently. (It is "plastic" rather than elastic.) A wooden plank will break if too great a force is exerted on its unsupported center. A rubber band will snap under too great a pull. A steel ball will be permanently flattened under too great a compression.

However, if one works with forces not strong enough to surpass this limit, one can arrive, as Hooke did, at a useful generalization that can be briefly expressed as follows:

The strain is proportional to the stress.

This is called *Hooke's law*. One would expect, from Hooke's law, that if a force x stretches a spring through a distance y, then a force of $2x$ will stretch it through a distance of $2y$, and a force of $x/2$ will stretch it through a distance of $y/2$. Suppose then that the amount of stretch produced by a known force is measured. Any force of unknown size (within the elastic limit) can then be measured by simply measuring the strain it produced.

This principle can be applied to any other form of stress that produces an easily measured strain; for instance, it can be applied to the twisting, or *torsion*, of an elastic rod or fiber. When torsion is used to measure the size of an unknown stress by the amount of strain produced, the set-up is called a *torsion balance*. If an extremely thin fiber is used, one that can be twisted by very small forces, it becomes conceivable that even tiny gravitational forces may be measured.

In 1798, the English scientist Henry Cavendish (1731–1810) made use of a delicate torsion balance for just this purpose. His torsion balance consisted of a light rod suspended at the middle by a delicate wire approximately a yard long. At each end of the light rod was a lead ball about two inches in diameter. Imagine a force applied to each lead ball in opposite directions and at right angles to both the rod and the delicate wire. The wire would twist, and extremely small forces would be sufficient to produce such a twist.

As a preparatory step Cavendish applied tiny forces to determine the amount of twist that would result. Next, carefully shielding his apparatus from air currents he brought two larger lead balls, each about eight inches in diameter, almost in contact with the small lead balls, but on opposite sides. The gravitational force between the lead balls now produced a twist in the fiber and from the total angle of twist Cavendish could measure the strength of the force between the small and large lead ball. (It turned out to be about 1/2,000,000 of a newton.)

Suppose, we rearrange Equation 4–1 as follows:

$$G = \frac{Fd^2}{mm'} \qquad \text{(Equation 4–2)}$$

With the value of F determined as I have just described, it is a simple matter to measure the mass of the lead balls (m and m') and the distance between them (d), center to center. Once all the values of the symbols on the right hand side of the equation are known, it is simple arithmetic to calculate the value of G. (Since the units of F in the mks system are kg-m/sec^2, those of d^2 are m^2, and those of mm' are kilograms times kilograms, or kg^2; the units

Cavendish experiment

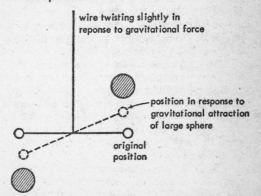

wire twisting slightly in reponse to gravitational force

position in response to gravitational attraction of large sphere

original position

of G work out, by Equation 4–2, to [(kg-m/sec²) m²]/kg², or m³/kg-sec².)

The best modern determination of G gives it a value of 0.0000000000667 m³/kg-sec², certainly a tiny enough value. It is a tribute to Cavendish's great talents as an experimenter that in his first determination he obtained a value very close to this.

Suppose, now, we arrange Equation 4–1 as follows:

$$m' = \frac{Fd^2}{Gm}$$ (Equation 4–3)

and try once more to determine the mass of the earth (m'). We already have, in the mks system, a value of 0.98 for F, one of 6,370,000 for d, and one of 0.1 for m. If we now add the value of 0.0000000000667 for G, it is simple arithmetic to solve for m', the mass of the earth. You can see that m' is equal to (0.98) (6,370,000)(6,370,000) divided by (0.0000000000667)(0.1), or just about 6,000,000,000,000,000,000,000,000 kilograms.

Physicists customarily express large numbers as powers of 10. Thus, 1,000,000 is usually written 10^6, which signifies the product of six 10's. The exponent (for numbers larger than 1) signifies the number of 0's in the number. It follows that 6,500,000 is 6.5×10^6. Negative exponents signify numbers less than 1, so that 10^{-6} is equal to $1 \cdot 10^6$, or 1/1,000,000 or 0.000001. Again, 0.00000235 is 2.35×10^{-6}.

Using such *exponential notation*, the value of G is 6.67×10^{-11} m³/kg-sec², and the mass of the earth is 6×10^{24} kg. (In the cgs system, the value of G is 6.67×10^{-8} cm³/gm-sec² and the mass of the earth is 6×10^{27} gm.)

CHAPTER 5

Weight

The Shape of the Earth

By determining the value of G, Cavendish had, in effect, determined the mass of the earth. For this reason, Cavendish is often said to have "weighed the earth," but this is *not* what he had done.

In common language, "weight" and "mass" are often spoken of as though they were the same things, and a body may be spoken of as "heavy" or "massive" interchangeably; even physicists sometimes fall into the trap. However, consider what weight is. The weight of a body is the force with which it is attracted to the earth. To repeat, weight is a force and has the units of a force!

A simple way of measuring the weight of an object is to suspend it from a coiled spring. In accordance with Hooke's law the force by which the body is attracted to the earth will extend the spring, the amount of extension (or strain) being proportional to the force (or stress). A weight-measuring device of this sort is a *spring balance*.

The mass of a body, on the other hand, is the quantity of inertia it possesses. By Newton's second law $m = f/a$; it is a force divided by an acceleration. Weight, which is a force, must by the same law be a mass multiplied by an acceleration. In the case of weight, which is the force of earth's gravitational field upon a body, the particular acceleration involved is, naturally, that produced by the earth's gravitational field.

The weight (w) of a body, in other words, is equal to the mass (m) of that body times the acceleration (g) due to the pull of earth's gravity:

$$w = mg \qquad \text{(Equation 5-1)}$$

Since the value of g is, under ordinary circumstances, just about constant, weight may be said to be directly proportional to mass. To say that A is 3.65 times as massive as B is equivalent to saying that under ordinary circumstances A is 3.65 times as heavy, or as weighty, as B. Since the two statements are usually equivalent, there is a strong temptation to consider them synonymous, and there lies the source of confusion between mass and weight.

The confusion is made worse because of the common units used for weight. A body with a mass of one kilogram is commonly said to have a weight of one kilogram, too.[*] In the mks system, however, the units of m are kilograms (kg), and the units of g are m/sec². Since weight is equal to mass times gravitational acceleration (mg), the units of weight are kg-m/sec² or newtons. A kilogram of mass therefore exerts (under ordinary circumstances) 9.8 newtons of force.

A kilogram of weight (which may be abbreviated as kg(wt.) to distinguish it from a kilogram of mass) is, therefore, *not* equal to 1 kg but to 9.8 newtons. In the cgs system, g is equal to 980 cm/sec². The weight of a body with a mass of 1 gm is therefore 1 gm multiplied by 980 cm/sec², or 980 gm-cm/sec². Consequently 1 gm (wt.) equals 980 dynes.

All this may strike you as unnecessarily puristic and refined —as making a great deal out of a distinction without a difference. After all, if weight and mass always vary in the same way, why bother so much about which is which?

The point is that mass and weight do not always vary in the same way. They are related by g, and the value of g is not a constant under all conditions.

The gravitational force (F) exerted by the earth upon a particular body is equal to mg, as indicated by Equation 5-1. It

_____

[*] The units of weight (pound, ounce, etc.) were in use long before Newton established the concept of mass. The units of weight were borrowed and applied to mass, which was a mistake—but one which is now beyond retrieval.

is also equal to Gmm'/d^2 as shown in Equation 4–1. Therefore $mg = Gmm'/d^2$; or dividing through by m:

$$g = \frac{Gm'}{d^2}$$

(Equation 5–2)

Of the quantities upon which the value of g depends in Equation 5–2, the gravitational constant (G) and the mass of the earth (m') may be considered as constant. The value of d, however, which is the distance of the body from the center of the earth is certainly not constant, and g varies inversely as the square of that distance.

An object at sea level, for instance, may be 6370 km from the center of the earth, but at the top of a nearby mountain it may be 6373 km from the center, and a stratoliner may take it to a height of 6385 km from the center.

Even if we confine ourselves to sea level, the distance to the center of the earth is not always the same. Under the action of gravity alone, the earth would be a perfect sphere (barring minor surface irregularities)—a fact pointed out by Aristotle—and then the distance from sea level to the earth's center would be the same everywhere. A second factor is introduced, however, by the fact that the earth rotates about its axis. This rotation means, as Newton was the first to recognize, that the earth cannot be a perfect sphere.

As the earth rotates about its axis, the surface of the earth is continually undergoing an acceleration inward toward the center of the earth (just as the moon does in revolving about the earth). If this is so, then Newton's third law (see page 34) comes into play. The earth's center exerts a constant force on the earth's outer layers to maintain that constant inward acceleration as the planet rotates; the outer layers must, therefore, by action and reaction, exert a force outward on the earth's center. The force directed inward is usually called a *centripetal force*, and the one directed outward is called a *centrifugal force* (the words coming from Latin phrases meaning "move toward the center" and "flee from the center," respectively).

The two forces are oppositely directed and the result is a stretching of the earth's substance. If you were to imagine a rope stretching from the earth's surface to the earth's center, with the earth's surfaces pulling outward at one end of the rope and the earth's center pulling inward at the other end, you would expect the rope to stretch by a certain amount; the earth's substance does exactly that.

If every point on the earth's surface were rotating at the same speed, the stretch would be the same everywhere and the earth would be perfectly spherical still. However, the earth rotates about an axis, and the nearer a particular portion of the earth's surface is to the axis, the more slowly it rotates. At the poles, the earth's surface touches the axis and the speed of rotation is zero. At the equator, the earth's surface is at a maximum distance from the axis and the speed of rotation is highest (just over 1600 kilometers an hour).

The interacting forces are zero at the poles, therefore, and increase smoothly as the equator is approached. The "stretch" increases, too, and a bulge appears in the earth, which reaches maximum size at the equator. Because of this equatorial bulge, the distance from the center of the earth to sea level at the equator is 21 km (13 miles) greater than the distance from the center of the earth to sea level at either pole.

The earth, therefore, is not a sphere, but an *oblate spheroid*.

To be sure, 21 km in a total distance of 6370 km is not much, but it is enough to introduce measurable differences in the value of *g*. What with the equatorial bulge and local differences in

Centrifugal force and equatorial bulge

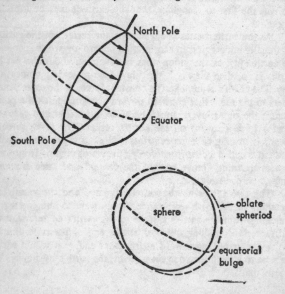

altitude, there are points in Alaska where the value of *g* is over 9.82 m/sec², whereas at the equator it is barely higher than 9.78 m/sec². That represents a difference of nearly one-half of one percent and is reflected in weight.

In other words, the weight of an object changes measurably from place to place on the earth's surface, as a spring balance would show. A man who weighs 200 pounds at the poles would weigh 199 pounds at the equator. To a chemist or physicist interested in the mass of an object (many properties depend on the mass), the measurement of weight as a substitute for mass would introduce serious inaccuracies.

What the scientist usually does when he "weighs" an object is to make use of a balance consisting of two pans suspended from opposite ends of a beam that is pivoted at the center. Objects of known weight are placed on one pan, the object to be weighed is placed in the other. The known weights are adjusted to the point of balance. The force of gravity is then the same on both pans; (if it were greater on one of them, that pan would sink downward while the other would rise upward).

If the weights are the same on both sides, then *mg* is the same on both sides. However much *g* might vary from point to point on the earth's surface, it is the same for two neighboring balance-pans at one particular point on the earth's surface. Therefore the mass (*m*) is the same for the objects in both pans. The mass of the unknown object is therefore equal to the mass of the weights (which is known).*

Beyond the Earth

Naturally, the small changes in the value of *g* become large ones if we alter greatly the distance of the body from the earth's center. A further complication is introduced if in removing the body to a great distance from the earth we bring it close to some other sizable conglomeration of mass. This situation is most likely to become important in connection with the moon, for man-made objects have already landed there, and living men may be standing on the surface of the moon before many years have passed.

An object on the moon's surface is still within the earth's gravitational field, which extends not only to the moon but, in principle, through all the universe. However, the moon also has

---

* Mass is not completely constant, by the way. However, the variation in mass becomes important only at extreme velocities not likely to be met with in ordinary life.

a gravitational field. That field is much smaller than the earth's, for the moon is much the less massive of the two. Nevertheless, an object on the moon's surface is much closer to the moon's center than to the earth's center; the moon's gravitational attraction would therefore be far greater than that of the distant earth, and a man standing on the surface of the moon would be conscious of the moon's pull only.

But the moon's pull on an object on its surface is not the same as the earth's pull on an object on its surface. To see how the two forces compare, let us refer back to Equation 4–1, which states that $F = Gmm'/d^2$. This F refers to the intensity of the earth's pull upon an object upon its surface. The moon's pull upon an object upon its surface, we can call F_m.

Now an object has the same mass whether it is on the surface of the earth or the surface of the moon, so m remains unchanged. The value of G is also unchanged, for it is constant throughout the universe. The mass of the moon, however, is known to be 1/81 the mass (m') of the earth. The mass of the moon, consequently, is $m'/81$. The distance from the surface of the moon to its center is 1737 km or just about 3/11 that of the 6370 km distance (d) from the surface of the earth to the center. Consequently, we can set the distance of the moon's surface to its center as $3d/11$.

We can now modify Equation 4–1, using the mass and radius of the moon to get an expression of the gravitational force of the moon for an object on its surface. This is:

$$F_m = \frac{Gm(m'/81)}{(3d/11)^2} \qquad \text{(Equation 5–3)}$$

If we now divide Equation 5–3 by Equation 4–1, we find that F_m/F (the ratio of the moon's gravitational force to that of the earth) is equal to 1/81 divided by $(3/11)^2$, or almost exactly 1/6. Thus, the gravitational force we would experience on the moon's surface would be 1/6 that to which we are accustomed on the surface of the earth. A 180-pound man who weighed himself on a spring balance would find he weighed 30 pounds.

But though the weight was decreased so drastically, the mass of an object would remain unchanged. This means that the force required to accelerate a particular object at a given rate would be the same on the moon as on the earth. We could lift a 180-pound friend without much trouble, for the sensation of the effort involved in the lift would be like that of lifting 30 pounds on earth. We could not, however, lift the man any more rapidly on the moon than on the earth. Now we could manipulate something that felt 30 pounds with a certain amount of ease on the earth. On the moon, something that felt 30 pounds would have six times the "normal" quantity of mass, and it could only be moved slowly. For this reason, maneuvering objects on the moon would give one the feel of "slow motion" or of pushing through molasses.

Again, if we jump on the moon, the force of our muscles will be countered by only 1/6 the gravitational force to which we are accustomed on the earth. The center of our body will therefore rise to six times the height it would on the earth. Having reached this unusual height, we would then fall toward the surface, but at 1/6 the usual acceleration (1.63 m/sec^2). This means we would seem to fall downward slowly and "like a feather." By the time we reached the surface again, however, having dropped at 1/6 the usual acceleration for six times the distance, we would be landing at just the same velocity that we would be landing at from a similar jump (equal exertion, but reaching a much lower height) on the earth.

To bring ourselves to a halt from that velocity would require as much force on the moon as it would on the earth, for it is mass that counts in this respect, not weight, and the mass remains unchanged on the moon. If we are seduced by our slow-paced fall into thinking we are indeed a feather, and try to land gracefully on one big toe, we are likely to break that toe.

The situation can be made even more radically unusual without having to go to the moon.

The subjective sensation we call "weight" arises from the fact that we are physically prevented from responding to the force of gravitation with an acceleration. Standing on the surface of the earth, we are prevented by the substance of the earth itself from an accelerated fall toward earth's center. It is the force exerted upon us against the stubborn opposition of the ground we stand on that we interpret as "weight."

If we were falling at precisely the acceleration imposed upon us by the gravitational acceleration (free fall), we would feel no weight. If we were in an elevator that had broken loose and was dropping without restraint, or if we were in an airplane that was in an unpowered fall, our sensation of weight would be gone. We could not press against the floor of the elevator or the airplane since that floor would be falling as rapidly as we were. If we were in midair within the elevator, we could not drop to its floor, for the floor would be moving as fast as we were. We would therefore seem to remain floating in midair and to be weightless.

Such examples of free fall are imperfect. Neither an elevator nor an airplane could fall for long without coming to disaster and ruining the experiment. Furthermore, the falling elevator or airplane would be slowed somewhat by the resistance of the air it was rushing through, and slowed to a greater extent than the man within the the elevator or airplane would be slowed by the quiet air about him. There would therefore be the sensation of some slight weight.

For the true feeling of free fall, we would need to be beyond the major portion of the atmosphere, say at a height of 160 km or more above the surface of the earth. To keep at that height it would be best if there were also a sideways motion that would keep one in orbit about the earth, in the same way that a combination of inward and sideways forces keeps the moon in orbit about the earth (see page 42).

This is exactly the situation in a manned orbiting satellite. Such a satellite is in free fall and can continue in free fall for long periods. The astronaut within has no sensation of weight. This is *not* because he is "beyond the pull of earth's gravity" as some news announcers maintain. It is only because he is in free fall, so the satellite and everything in it are falling at precisely the same acceleration.

The earth itself is in free fall in an orbit that takes it around the sun. Although its mass is huge (see page 52), its weight is zero. Cavendish, therefore, did not "weigh the earth," for he did not need to; its weight was understood to be zero from Newton's time. What Cavendish did was to determine the earth's mass.

Even in free fall, where weight is zero, the mass of a particular body remains unchanged. Astonauts building a space station will be moving huge girders that will have no weight. They will be able to balance such girders on one finger, if girder and finger are motionless with respect to each other. If a girder must be set in motion, however, or if it is already moving and must either be stopped or have its direction of motion altered, the effort will be precisely as great as it would be on the earth. A man trapped between two girders moving toward each other may well be crushed to death by two weightless but not massless objects.

The distinction between mass and weight, which seems so trivial on the surface of the earth, is therefore anything but trivial in space, and can easily become a matter of life and death.

Escape Velocity

As an object is dropped from a greater and greater height above the ground, it takes longer and longer to fall and strikes the ground with a higher and higher velocity. If we use Equations 2–1 and 2–2 (see page 17-18), letting the symbol a in those equations be set equal to 9.8 m/sec^2 (the acceleration in free fall), we can make some easy calculations. A body dropped from a height of 4.9 meters will strike the ground in one second and be moving, at the moment of impact, at 9.8 m/sec. If it were dropped from a height of 19.6 meters, it would strike after two seconds, moving then at 19.6 m/sec. If it were dropped from 44.1 meters, it would strike after three seconds, moving then at 29.4 m/sec.

It would seem that if you could only drop an object from a great enough height, you could make the velocity of impact as high as you pleased. Certainly this would seem so if the value of g were the same for all heights.

But the value of g is not constant; it decreases with height. The value of g varies inversely as the square of the distance from the earth's center. A point 6370 kilometers above the earth's surface would be 12,740 kilometers from the earth's center—twice as far from the center as a point on the surface would be. The value of g at that height would therefore be just 1/4 what it is at the surface.

An object falling from an initial state of rest 6370 kilometers above the earth's surface would in the first second attain a velocity of only 2.45 m/sec, instead of the 9.8 m/sec it would attain after a one-second drop in the immediate vicinity of the earth's surface.

As the body continued to drop and approach the earth, the value of g would, of course, increase steadily and approach 9.8 m/sec^2 at the end. However, the falling body would not strike the earth's surface with as high a velocity of impact as it would have done if the value of g had been 9.8 m/sec^2 all the way down.

Imagine a body dropped first from a height of 1000 kilometers, then from 2000 kilometers, then from 3000 kilometers, and so on. The drop from 1000 kilometers would result in a velocity of impact, v_1. If the value of g were constant all the way up, then a drop from 2000 kilometers would involve a gain in velocity in the first 1000 kilometers equal to the gain in the second 1000 kilometers, so the final velocity of impact would be $v_1 + v_1$ or $2v_1$. However, the upper 1000 kilometers represents a region where g is smaller than in the lower 1000 kilometers. Less velocity is added in the upper than in the lower half of the drop, and the final velocity of impact is $v_1 + v_2$, where v_2 is smaller than v_1. The same argument can be repeated all the way up, so a fall from a height of 10,000 kilometers would result in a velocity of impact of $v_1 + v_2 + v_3 + v_4$ and so on, up to v_{10}. Here each symbol represents the portion of the final velocity contributed by a higher and higher 1000 kilometer region of drop, and the value of each symbol is less than that of the preceding one.

Whenever one is faced with a series of numbers each smaller than the one before, there is the possibility of a *converging series*. In such a series, the sum of the numbers never surpasses a certain fixed value, the *limiting sum*, no matter how many numbers are added. The best-known case of such a converging series is $1 + 1/2 + 1/4 + 1/8 + 1/16$, where each number is half the one before. The sum of the first two numbers is 1.5; the sum of the first three numbers is 1.75; the sum of the first four numbers is 1.875; the sum of the first five numbers is 1.9325, and so on. As more and more numbers in the series are added, the sum grows larger and larger, and approaches closer and closer to 2 without ever quite reaching it. The limiting sum of this particular series is 2.

It turns out that the numbers representing increments of velocity resulting from falls from regularly increased heights do indeed form a converging series. As a body is dropped from a greater and greater height, the final velocity of impact does not increase without limit; instead it tends toward a limiting velocity it cannot surpass.

This limiting velocity of impact (v_l) depends on the value of g and on the radius (r) of the body that is the source of the gravitational field. The importance of the radius rests on the fact that the larger its value, the more slowly does the value of g fade off with distance. Suppose a body has a radius of 1000 kilometers. At 10,000 kilometers from its center, a falling body is ten times as far from the center as an object on the surface is, and the value of g is therefore 1/100 the value at the surface. Suppose that a body has a radius of 2000 kilometers, however; at a distance of 10,000 kilometers from its center, a falling body is then only five times as far from the center as an object on the surface is, and the value of g is 1/25 the value at the surface. Through all heights, therefore, the value of g would decline more rapidly for the small body than for the large body, and the final velocity of impact would be less for the smaller body even though its surface value of g might be the same as for the larger body.

It turns out that:

$$v_1 = \sqrt{2gr} \qquad \text{(Equation 5-4)}*$$

In the mks system, the value of g is 9.8 m/sec² and that of r is 6,370,000 m, so that $2gr$ is equal to about 124,800,000 m²/sec². In taking the square root of this number, we must also take the square root of the unit. Since the square root of a^2/b^2 is ab, it should be clear that the square root of m²/sec² is m/sec. The square root of 124,800,000 m²/sec² is about 11,200 m/sec. This limiting velocity of impact is equal to 11.2 km/sec (or just about seven miles per second). No object, falling to earth *from rest*, could ever strike with an impact of more than 11.2 km/sec. (Of course, if an object such as a meteor happens to be speeding in the direction of the earth, so that its own speed is added to the velocity produced by earth's gravitational field, it will strike with an impact of more than 11.2 km/sec.)

For the moon, with its smaller values for both g and r, the maximum velocity of impact is only 2.4 km/sec (or 1.5 miles per second).

Suppose we now turn the matter around. Instead of a falling body, consider one that is propelled upward. For a body moving upward, g represents the amount by which its speed is diminished each second (see page 29). The situation is symmetrically reversed to that of a body moving downward; that is, if a body initially at rest falls from a height h and attains a velocity v at the moment of impact, then a body hurled upward with a velocity v will attain a height h before coming to rest (and beginning its fall back to the earth).

But a body dropped from any height, however great, can never attain a velocity of impact greater than 11.2 km/sec. This means that if a body is hurled upward with a velocity of 11.2 km/sec or more it will never reach a point of rest and, therefore, never fall back to the earth (barring the interference of gravitational fields of other bodies).

The limiting velocity of impact is consequently also the velocity at which a body hurled upward will escape from the earth permanently; it is therefore called the *escape velocity*. The escape velocity at the surface of the earth is 11.2 km/sec and the escape velocity at the surface of the moon is 2.4 km/sec.

---

* As this is an introduction to physics, I shall not always give the derivation of the equations used—since at times these would involve concepts not yet explained or mathematical techniques I would prefer not to use.

A body that orbits the earth has not escaped from the earth. It is falling toward it, and only its sideward motion prevents it from finally making impact. A smaller velocity is therefore required to place an object in orbit than to cause it to escape from the earth altogether. For a circular orbit, the velocity must be equal to $\sqrt{gr}$, where r is the distance of the orbiting body from the earth's center and g is the value of the gravitational acceleration at that distance. In the immediate vicinity of the earth's surface, this comes out to 7.9 km/sec (or 4.9 miles per second). Orbiting satellites travel at this velocity and complete the 40,000 kilometer circumnavigation of the earth in a minimum time of 85 minutes.

As the distance from the earth's center increases, the value of r increases, of course, while the value of g decreases, varying as $1/r^2$. The variation of $\sqrt{gr}$ (which is the orbital velocity) as distance increases is as $\sqrt{(1/r^2)(r)}$ or $\sqrt{1/r}$. In other words, the orbital velocity of a body varies inversely as the square root of its distance from the object around which it is in orbit.

Thus, the distance of the moon from the earth's center is 382,400 kilometers. This is 60.3 times the distance from the center of a satellite orbiting just above the atmosphere. The orbital velocity of the moon is therefore less than that of the satellite by a factor equal to the $\sqrt{60.3}$. The moon's orbital velocity, in other words, is $7.9/\sqrt{60.3}$, or just about 1 km/sec.

Consider, also, a satellite in orbit 42,000 kilometers from the earth's center (about 35,600 kilometers above its surface). Its distance from the earth's center would be 6.6 times that of an object on earth's surface. Its orbital velocity would therefore be $7.9/\sqrt{6.6}$, or not quite 3.1 km/sec. The length of its orbit would be about 264,000 kilometers, and at its orbital velocity it would take the satellite just 24 hours to complete one revolution. It would, therefore, just keep pace with the surface of the rotating earth and would seem to hang motionless in the sky. Such apparently motionless satellites serve admirably as communication relays.

CHAPTER 6

Momentum

Impulse

Let's consider a falling body again.

An object held at some point above the ground is at rest. If it is released, it begins to fall at once. Motion is apparently created where it did not previously exist. But the word "created" is a difficult one for physicists (or for that matter philosophers) to swallow. Can anything really be created out of nothing? Or is one thing merely changed into a second, so the second comes into existence only at the expense of the passing into nonexistence of the first? Or perhaps one object undergoes a change (from rest to motion, for instance) because, and only because, another object undergoes an opposing change (from rest to motion in the opposite direction, for instance). In this last case, what is created is not motion but motion plus "anti-motion," and if the two together cancel out to zero, there is perhaps no true creation at all.

To straighten this matter out, let's start by trying to decide exactly what we mean by motion.

We can begin by saying that a force certainly seems to create motion. Applied to any body initially at rest, say to a hockey puck on ice, a force initiates an acceleration and sets the puck moving faster and faster. The longer the force acts, the faster the hockey puck moves. If the force is constant, then the velocity at any given time is proportional to the amount of the force multiplied by the time during which it is applied. The term *impulse* (I) is applied to this product of force (f) and time (t):

$$I = ft \qquad \text{(Equation 6–1)}$$

Since a force produces motion, we might expect that a given impulse (that is, a given force acting over a given time) would always produce the same amount of motion. If this is so, however, then the amount of motion cannot be considered a matter of velocity alone. If the same force acts upon a second hockey puck ten times as massive as the first, it will produce a smaller acceleration and in a given time will bring about a smaller velocity than in the first case. The quantity of motion produced by an impulse must therefore involve mass as well as velocity.

That this is indeed so is actually implied by Equation 6–1. By Newton's second law we know that a force is equal to mass times acceleration ($f = ma$). We can therefore substitute ma for f in Equation 6–1 and write:

$$I = mat \qquad \text{(Equation 6–2)}$$

But by Equation 2–1 (see page 17), we know that for any body starting at rest the velocity (v) produced by a force is equal to the acceleration (a) multiplied by time (t), so that $at = v$. If we substitute v for at in Equation 6–2, we have:

$$I = mv \qquad \text{(Equation 6–3)}$$

It is this quantity, mv, mass times velocity, that is really the measure of the motion of a body. A body moving rapidly requires a greater effort to stop it than does the same body moving slowly. The increase in velocity adds to its total motion, therefore. On the other hand, a massive body moving at a certain velocity requires a greater effort to stop it than does a light body moving at the same velocity. The increase in mass also adds to total motion. Consequently, the product mv has come to be called *momentum* (from a Latin word for "motion").

Equation 6–3 means that an impulse (ft) applied to a body at rest causes that body to gain a momentum (mv) equal to the impulse. More generally, if the body is already in movement, the application of an impulse brings about a change of momentum equal to the impulse. In brief, impulse equals change of momentum.

The units of impulse must be those of force multiplied by those of time, according to Equation 6–1, or those of mass multiplied by those of velocity, according to Equation 6–3. In the mks system, the units of force are newtons, so impulse may be measured in newton-sec. The units of mass are kilograms, however, and the units of velocity are meters per second, so the units of impulse (mass times velocity) are kg-m/sec. However, a newton has been defined as a kg-m/sec². A newton-sec, therefore, is a kg-m-sec/sec², or a kg-m/sec. Thus the units of I considered as ft are the same as the units of I considered as mv. In the cgs system, it is easy to show, the units of impulse are dyne-sec, or gm-cm/sec, and these are identical also.

Conservation of Momentum

Imagine a hockey puck of mass m speeding across the ice at a velocity, v. Its momentum is mv. Imagine another hockey puck of the same mass moving at the same speed but in the opposite direction. Its velocity is therefore $-v$ and its momentum is $-mv$. Momentum, you see, is a vector, since it involves velocity, and not only has quantity but direction. Naturally, if we have two bodies with momenta in opposite directions, we can set one momentum equal to some positive value and the other equal to some negative value.

Suppose now that the two hockey pucks are rimmed with a layer of glue powerful enough to make them instantly stick together on contact. And suppose they do make contact head-on. When that happens, they would come to an instant halt.

Has the momentum been destroyed? Not at all. The total momentum of the system* was $mv + (-mv)$, or 0, before the collision and $0+0$, or (still) 0, after the collision. The momentum was distributed among the parts of the system differently before and after the collision, but the total momentum remained unchanged.

Suppose that instead of sticking when they collided (an inelastic collision) the two pucks bounced with perfect springiness (an elastic collision). It would then happen that each puck would reverse directions. The one with the momentum mv would now have the momentum $-mv$ and vice versa. Instead of the sum $mv+(-mv)$, we would have the sum $(-mv)+mv$. Again there would be a change in the distribution of momentum, but again the total momentum of the system would be unchanged.

---

* By a "system" is meant the entire collection of bodies being discussed, in this case, the two hockey pucks, considered in isolation from the rest of the universe.

If the collision were neither perfectly elastic nor completely inelastic, if the pucks bounced apart but only feebly, one puck might change from mv to $-0.2mv$, while the other changed from $-mv$ to $0.2mv$. The final sum would still be zero.

This would still hold true if the pucks met at an angle, rather than head-on, and bounced glancingly. If they met at an angle, so their velocities were not in exactly opposite directions, the two momenta would not add up to zero, even though the velocities of the two pucks were equal. Instead the total momentum of the system would be arrived at by vector addition of the two individual momenta. The two pucks would then bounce in such a way that the vector addition of the two momenta after the collision would yield the same total momentum as before. This would also be true if a moving puck struck a puck at rest a glancing blow. The puck at rest would be placed in motion, and the originally moving puck would change its direction; however, the two final momenta would add up to the original.

Matters would remain essentially unchanged even if the two pucks were of different masses. Suppose one puck was moving to the right at a given speed and had a momentum of mv, while an-

Conservation of momentum

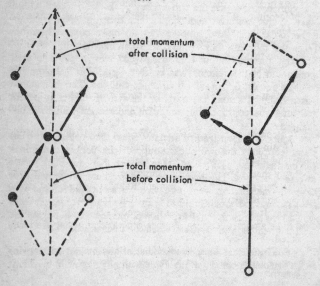

other, three times as massive, was moving at the same speed to the left and had, therefore, a speed of $-3mv$. If the two stuck together after a head-on collision, the combined pucks (with a total mass of $4m$) would continue moving to the left—the direction in which the more massive puck had been moving—but at half the original velocity ($-v/2$). The original momentum of the system was $mv + (-3mv)$, or $-2mv$. The final momentum of the system was $(4m)(-v/2)$, or $-2mv$. Again, the total momentum of the system would be unchanged.

And what if momentum is seemingly created? Let us consider a bullet initially at rest—and with a momentum, therefore, of 0—which is suddenly fired out of a gun and moves to the right at high velocity. It now has considerable momentum (mv). However, the bullet is only part of the system. The remainder of the system, the gun, must gain $-mv$ by moving in the opposite direction. If the gun has n times the mass of the bullet, it must move in the opposite direction with $1/n$ times the velocity of the speeding bullet. The momentum of the gun (minus the bullet) would then be $(nm)(-v/n)$, or $-mv$. (If the gun were suspended freely when it was fired, its backward jerk would be clearly visible. When fired in the usual manner its backward motion is felt as "recoil.") The total momentum of gun plus bullet was therefore 0 before the gun was fired and 0 after it was fired, though here the distribution of momentum among the parts of the system varied quite a bit before and after firing.

In short, all the experiments we can make will bring us to the conclusion that:

The total momentum of an isolated system of bodies remains constant.

This is called the *law of conservation of momentum*. (Something that is "conserved" is protected, guarded, or kept safe from loss.)

Of course, it is impossible to prove a generalization by merely enumerating isolated instances. No matter how often you experiment and find that momentum is conserved, you cannot state with certainty that it will *always* be conserved. At best, one can only say, as experiment after experiment follows the law and as no experiment is found to contradict it, that the law is increasingly probable. It would be far better if one could show the generalization to be a consequence of another generalization that is already accepted.

For instance, suppose two bodies of any masses and moving at any velocities collide at any angle with any degree of elasticity.

At the moment of collision, one body exerts a force (f) on the second. By Newton's third law, the second body exerts an equal and opposite force ($-f$) on the first. The force is exerted only while the two bodies remain in contact. The time (t) of contact is obviously the same for both bodies, for when the first is no longer in contact with the second, the second is no longer in contact with the first. This means that the impulse of the first body on the second is ft, and that of the second on·the first is $-ft$.

The impulse of the first body on the second imparts a change in momentum mv to the second body. But the impulse of the second body on the first, being exactly equal in quantity but opposite in sign, must impart a change in momentum $-mv$ to the first. The changes in momentum may be large or small depending on the size of the impulse, the angle of collision, and the elasticity of the material; however, whatever the change in momentum of one, the change in the other is equal in size and opposition in direction. The total momentum of the system must remain the same.

Thus, the law of conservation of momentum can be derived from Newton's third law of motion. In actual fact, however, it was not, for the law of conservation of momentum was first enunciated by an English mathematician,·John Wallis (1616–1703), in 1671, a dozen years before Newton published his laws of motion. One could, indeed, work it the other way, and derive the third law of motion from the law of conservation of momentum.

At·this point you might feel that if the physicist proves the conservation of momentum from the third law of motion, and then proves the third law of motion from the conservation of momentum, he is actually arguing in a circle and not proving anything at all. He would be if that were what he is doing, but he is not.

It is not so much a matter of "proving" as of making an assumption and demonstrating a consequence. One can begin by assuming the third law of motion and then showing that the law of conservation of momentum is a consequence of it. Or one can begin by assuming the law of conservation of momentum and showing that the third law is the consequence of that.

The direction in which you move is merely a matter of convenience. In either case, no "proof" is involved and no necessary "truth." The whole structure rests on the fact that no one in nearly three centuries has been able to produce a clearcut demonstration that a system exists, or can be prepared, in which either the third law of motion or the law of conservation of momentum is not obeyed. Such a demonstration may be made tomorrow, and the foundations of physics may have to be modified as a consequence; but by now it seems very unlikely* that this will happen.

And yet it may be that with a little thought we might think of cases where the law is not obeyed. For instance, suppose a billiard ball hits the rim of the billiard table squarely and rebounds along its own line of approach. Its velocity v becomes $-v$ after the rebound, and since its mass remains unchanged, its original momentum mv has become $-mv$. Isn't that a clear change in momentum?

Yes, it is, but the billiard ball does not represent the entire system. The entire system includes the billiard table that exerted the impulse that altered the billiard ball's momentum. Indeed, since the billiard table is fixed to the ground by frictional forces too large for the impact of the billiard ball to overcome, it includes the entire planet. The momentum of the earth changes just enough to compensate for the change in the momentum of the billiard ball. However, the mass of the earth is vastly larger than that of the billiard ball, and its change in velocity is therefore correspondingly smaller—far too small to detect by any means known to man.

Yet one might assume that if enough billiard balls going in the same direction were bumped into enough billiard tables, at long, long last, the motion of the earth would be perceptibly changed. Not at all! Each rebounding billiard ball must strike the opposite rim of the table, or your hand, or some obstacle. Even if it comes to a slow halt through friction, that will be like striking the cloth of the table little by little. No matter how the billiard ball moves it will have distributed its changes in momentum equally in both directions before it comes to a halt, if only itself and the earth are involved.

A more general way of putting it is that the distribution of momentum among the earth and all the movable objects on or near its surface may vary from time to time, but the total momentum, and therefore the net velocity of the earth *plus* all those movable objects (assuming the total mass to remain constant), must remain the same. No amount or kind of interaction among the components of a system can alter the total momentum of that system.

And now the solution to the problem of the falling body with which I opened the chapter is at hand. As the body falls it gains momentum (mv), this momentum increasing as the velocity increases. The system, however, does not consist of the falling body

---

* Please remember that "unlikely" does not mean "impossible."

alone. The gravitational force that brings about the motion involves both the body and the earth. Consequently, the earth must gain momentum ($-mv$) by rising to meet the body. Because of the earth's huge mass, its upward acceleration is vanishingly small and can be ignored in any practical calculation. Nevertheless, the principle remains. Motion is not created out of nothing when a body falls. Rather, both the motion of the body and the anti-motion of the earth are produced, and the two cancel each other out. The total momentum of earth and falling body, with respect to each other, is zero before the body starts falling, is zero after it completes its fall, and is zero at every instant during its fall.

Rotational Motion

So far, I have discussed motion as though it involved the displacement of an object as a whole through space with the different parts of the object maintaining their mutual orientation unchanged. This is *translational motion* (from Latin words meaning "to carry across").

It is possible, however, for a body not to be displaced through space as a whole, and yet still be moving. Thus, the center of a wheel may be fixed in place so that the wheel as a whole does not change its position; nevertheless, the wheel may be spinning about that center. In similar fashion, a sphere fixed within a certain volume of space may yet spin about a fixed line, the axis. This kind of motion is *rotational motion* (from the Latin word for "wheel"). (It is, of course, possible for a body to move in a combination of these two types of motion, as when a baseball spins as it moves forward, or when the earth rotates about its axis as it moves forward in its orbit about the sun.)

Rotational motion is quite analogous to translational motion, but it requires a change of viewpoint. For instance, we are quite used to thinking of the speed of translational motion in terms of miles per hour or centimeters per second. Furthermore, we take it for granted that if one part of a body has a certain translational velocity so has every other part of the body. The tail of an airplane, in other words, moves forward just as rapidly as its nose.

In the case of rotational motion, matters are different. A point on the rim of a turning wheel is moving at a certain speed, a point closer to the center of the wheel is moving at a smaller speed, and a point still closer to the center is moving at a still smaller speed. The precise center of a turning wheel is motion-less. To say that a turning wheel moves at so many centimeters per second is therefore meaningless, unless we specify the exact portion of the wheel to which we refer, and this can be most incon-venient.

It would be neater if we could find some method of measuring rotational speed that would apply to the entire rotating body at once. One method might be to speak of the number of revolu-tions in a unit time. Though various points on the wheel might move at various speeds, every point on the wheel completes a revolution in precisely the same period, since the wheel rotates "all in one piece." We might therefore speak of a wheel or any rotating object as having a speed of so many *revolutions per minute* (usually abbreviated as *rpm*).

Or we might divide one revolution into 360 equal parts called *degrees* and abbreviated as a zero superscript (°). In that case 1 rpm would be equal to 360° per minute, or 6° per second. As the wheel sweeps out those degrees, a line connecting the center of the wheel with a point on its rim marks out an angle. A speed given in revolutions per minute or degrees per second is therefore spoken of as *angular speed*.

It is possible for rotational motion to take place in one of two mirror-image fashions. As viewed from a fixed position, a wheel may be observed to be rotating *clockwise*—that is, in the same sense that the hands of a clock move. It could, on the other hand, move *counterclockwise*—that is, in the opposite sense to the moving clockhand.* Therefore, we can speak of *angular velocity* as indicating not only speed but direction as well. (Veloci-ties involved in translational motion can be spoken of as *linear velocity*, since movement is then along a line rather than through an angle.)

Physicists use another unit in measuring rotational velocity: the *radian*. This is an angle that marks out on the rim of a circle an arc that is just equal in length to the radius of the circle. The circumference of the circle is π times the length of the

---

* It is important to specify "from a fixed position," for clockwise and counterclockwise are not absolute terms. A wheel may seem to be turning clockwise to you, but if you move to the opposite side and view it, it will then seem to be turning counterclockwise. The same is true if you speak of transla-tional motion as being "left" or "right," or "toward" or "away." Those are terms that have meaning only with reference to your own position. However, if you speak of "north," "south," "east," or "west," those are terms that are fixed with respect to the earth and do not depend on your own position.

diameter* and is therefore 2π times the length of the radius. The circumference must therefore also be 2π times the length of the arc marked out by one radian. The entire circumference is marked out in one revolution, so one revolution equals 2π radians or 360°. It follows that one radian equals $360°/2\pi$ or, since π equals 3.14159, one radian is about equal to 57.3°.

Angular velocity is often symbolized by the Greek letter ω ("omega"), since this is the equivalent in Greek of the Latin letter v usually used for linear velocity.

For any given point on a rotating body, angular velocity can be converted to linear velocity. The linear velocity depends not only on the angular velocity but also on the distance (r) of the point in question from the center of rotation. If the distance is doubled for the same angular velocity, the linear velocity of the point is doubled. We can say then that:

$$v = r\omega \qquad\qquad \text{(Equation 6-4)}$$

This equation is precisely correct when ω is measured in radians per unit of time. For instance, if the angular velocity is one radian per second, then in one second a given point anywhere on the wheel sweeps out an arc equal to its distance from the center, and $v=r$. If ω equals 2 radians per second then $v=2r$, and so on.

If we were measuring ω in revolutions per unit of time, then Equation 6-4 would have to read $v = 2\pi r\omega$, and if we were

---

* The Greek letter π ("pi") is used to represent the ratio of the circumference (c) of a circle to its diameter (d); in other words, $c/d = \pi$. Although every circle may have a different value for c and for d, the ratio of the two, c/d, is always the same for all circles. Therefore, π is a constant, and it is approximately equal to 3.14159.

Size of the radian

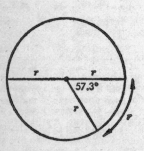

Angular velocity

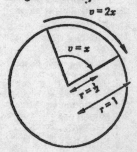

measuring it in degrees per unit time, it would have to read $v = r\omega/57.3$. This is an example of how a unit that may, at first blush, seem to have an odd and inconvenient size can yet turn out to be useful because it allows relationships to be expressed with maximum simplicity.

Torque

It takes a force to set a body at rest into translational motion. Under certain conditions a force can set a body at rest into rotational motion instead. Suppose, for instance, you nailed a long, flat rod loosely to a wooden base at one end. If you pushed the rod, it would not move as a whole, in a translational manner, because one end is fixed. The rod would instead begin to make a rotational movement about the fixed end.

A force that gives rise to such a rotational movement is called a *torque* (from a Latin word meaning "to twist"). To continue the use of Greek letters for rotational motion, a torque may be symbolized by the Greek letter τ ("tau"), which is the equivalent of the Latin *t* (for "torque," obviously).

A given force does not always give rise to the same torque by any means. In the case of the rod just mentioned, the amount of torque depends on the distance from the point at which the force is applied to the fixed point. A force applied to the fixed point will not itself produce a torque. As one recedes from that point a given force will produce a more and more rapid rotation and will therefore represent a greater and greater torque. In fact, the torque is equal to the force (f) multiplied by the distance (r):

$$\tau = fr \qquad \text{(Equation 6–5)}$$

In the past, torque has been referred to as the *moment of force*, but this phrase is now quite out of fashion.

Nor need a torque be produced only where a portion of a body is fixed in space; one can be produced even if the entire body is free to move.

Consider a body possessing mass but consisting of but a single point. Such a body can only undergo translational motion. A rotating body, after all, spins about some point (or line); if that point is all that exists, then there is nothing to spin and only linear motion is possible. It is to such point-masses that the laws of motion can be made to apply most simply.

In the real universe, however, there are no point-masses. All real massive bodies have extension. Nevertheless, it can be shown that in some ways such real bodies behave as if all their mass were concentrated at some one point. The point at which this seeming concentration is found is the *center of mass*. Where a body is symmetrical in shape, and where it is either uniform in density or has a density that changes in symmetrical fashion, the center of mass is at the geometrical center of the body. For instance, the earth is an essentially spherical body; while it is not uniformly dense, it is most dense at the center, and this density falls off equally in all directions as one approaches the surface. The earth's center of mass therefore coincides with its geometric center, and it is toward that center that the force of gravity is directed.

The concept of the center of mass can explain several things that might otherwise be puzzling. According to Newton's first law of motion, a moving object will continue moving at constant velocity unless acted upon by some outside force. Suppose that a shell containing an explosive is moving at constant velocity through space and that at a certain point it explodes. Fragments of the shell are hurled in all directions, and the various chemical products of the explosion also expand outward. This explosion is an internal force, however, one produced within the system in question, and it should have no effect on the motion of the system, according to the first law. Yet the various fragments of the shell are no longer traveling at the original velocity. Do Newton's laws of motion break down?

Not at all. The laws apply to a system as a whole, and not necessarily to one part or another taken in isolation. As a result of the explosion the system has changed its shape. But has it changed its center of mass? The center of mass might be viewed as the "average point" of the body. If one portion of the shell hurls outward in one direction, it is balanced by another portion hurled in the opposite direction. To be more precise, the vector sum of all the momenta in one direction must be equal to the vector sum of all the momenta in the opposite direction, according to the law of conservation of momentum. This can be shown to imply that no matter how the body changes shape through internal forces, the center of mass remains where it would have been if no change of shape had occurred. In other words, the center of mass of the system moves on at constant velocity regardless of the explosion that hurled bits of the system this way and that.

If a body were under the influence of a gravitational force

and following a parabolic path, its sudden explosion would not prevent the center of mass from continuing smoothly in that parabolic path even though the individual fragments moved all over the lot. (This implies no interference by forces outside the system. If fragments strike other bodies and are halted, the motion of the center of mass changes. Again, the effect of air resistance on the multitude of particles after explosion may not be the same as the effect upon the single shell before explosion; this may change the motion of the center of mass.)

Suppose next that a body is falling toward the earth. Every particle of the body is being pulled by the force of gravity, but the body behaves as if all that force were concentrated at one point within the body; that point is the *center of gravity*. If the body were in a uniform gravitational field, the center of gravity would be identical with the center of mass. However, the lower portion of a body is somewhat closer to the center of the earth than is the upper, and the lower portion is therefore more strongly under gravitational influence. The center of gravity is consequently very slightly below the center of mass; therefore, while the difference under ordinary conditions is so small as to be easily neglected, it is better form not to interchange the two phrases.

The concept of the center of gravity is useful in considering the stability of bodies. Imagine a brick resting on its narrowest base. If it is tipped slightly and then released, it drops back to

Center of gravity

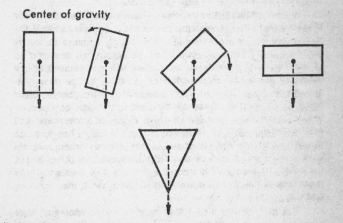

its original position. If it is tipped somewhat more and then re-leased, it drops back again. As it is tipped more and more, how-ever, there comes a point when it flops over onto one of its other bases. At what point does this flop-over come?

We can view the gravitational force as pulling upon the center of gravity of the brick, and upon that point only. As long as the center of gravity is located directly over some portion of the original base, the effect of the gravitational pull is to move the brick back upon that base once the tipping force is removed. If the brick is tipped so much that the center of gravity is located directly over some point outside the original base, the brick drops onto the base over which the point is now located.

Naturally, the wider the base in comparison with the height of the center of gravity, the greater the degree of tipping required before the center of gravity moves beyond that base—the more stable the body, in other words. A brick resting on its broadest base is more stable than one resting upon its narrowest base.

A cone resting on its pointed end may be so adjusted that its center of gravity will be directly above that point. It will then remain in balance. The slightest movement, however, the smallest breath of air, will move the center of gravity beyond the point in one direction or another, and down it will flop onto its side. A juggler keeps objects balanced upon points or, more accurately, upon very small bases, by moving his own body in such a way as to bring the base under the center of gravity again every time the center of gravity moves out of position.

Where a body is not uniform in density, its center of gravity is not located at its geometrical center but is displaced toward the denser portions. An object that is particularly dense in its lower-most portion ("bottom-heavy") has an unusually low center of gravity. Even a large degree of tipping will not bring that low center of gravity beyond the line of the base, and on being re-leased the object will return to its original position. On the other hand, an object that is particularly dense in its uppermost portion ("top-heavy") has an unusually high center of gravity and will flop over after but a slight degree of tipping. Since our common experience is with objects of reasonably uniform density, we are generally surprised at the refusal of a bottom-heavy object to fall over (like those round-bottomed children's toys that spring up again even when forced down to their side), or at the ease with which a top-heavy object topples.

Let us now return to our point-mass which undergoes only translational motion. If we imagine a force directed against a real body in such a way as to intersect its center of mass, then that real body behaves as a point-mass does and undergoes a purely translational motion. Thus, a falling body is subjected to the pull of gravity directly upon the center of gravity (usually equivalent to the center of mass). Therefore (provided a torque is not applied to the body at the moment of its release and that the effect of wind and air resistance is neglected), a body will fall in a purely translational manner.

If, however, a force is applied to a body in such a way as to be directed to one side or another of the center of mass, a torque is produced as well. Such bodies, even when forced into translational motion by the force, undergo rotational motion also. The manner in which footballs, baseballs and similar objects spin as they move is well known to all of us. It is so difficult to center the force upon the center of mass that it is virtually impossible to keep them from spinning.

Naturally, the further the force from the center of mass, the greater is the rotational motion compared with the translational motion. A coin standing on edge can be flicked into a spin by snapping a finger against its rim, and it will turn very rapidly while moving forward only slowly.

It is thought that the stars and planets originated by accretion and that small fragments struck the growing nuclei of the bodies. Astronomers have labored to devise schemes whereby these colliding bodies would be shown as tending to strike more often to one side of the center of mass than the other, setting up torques that do not average out to zero. Thus heavenly bodies, however they may move translationally, also rotate about some axis.

Conservation of Angular Momentum

There is rotational inertia as well as the more familiar translational inertia. If a wheel is rotating about a frictionless axis, it will continue to rotate at a constant angular velocity unless an outside torque is exerted upon it.

The application of a torque will induce an acceleration in the angular motion. This angular acceleration can be represented by α, the Greek letter "alpha," which is the equivalent of the Latin a. The units of angular acceleration are radians per second per second or radians/sec^2. Just as linear velocity is equal to angular velocity times distance from the center of rotation (see Equation 6-4) so, by the same line of reasoning, linear acceleration (a) is equal to angular acceleration (α) times distance from the center (r), or:

$$a = r\alpha \qquad \text{(Equation 6–6)}$$

By the second law of motion, we know that force is equal to mass times linear acceleration ($f = ma$). Combining this with Equation 6–6, we can substitute $r\alpha$ for a, and have:

$$f = mr\alpha \qquad \text{(Equation 6–7)}$$

We have already decided that torque (τ) is equal to force times distance from the center (fr). This was expressed in Equation 6–5 on page 75. Substituting the value for f given in Equation 6–7, we have:

$$\tau = (mr\alpha)\,(r) = mr^2\alpha \qquad \text{(Equation 6–8)}$$

Now according to the laws of motion as applied to translation, the ratio of the force to the acceleration (f/a) is the mass (m) (see Equation 3–3, on page 33). What if we take the analogous ratio in angular motion—that is, the ratio of the torque to the angular acceleration (τ/α)? By rearranging Equation 6–8, we can obtain a value for such a ratio:

$$\frac{\tau}{\alpha} = mr^2 \qquad \text{(Equation 6–9)}$$

In rotational motion, therefore, the quantity mr^2 (mass times the square of the distance from the center of rotation) is analogous to mass alone (m) in translational motion. This introduces interesting differences in the two types of motion.

Consider a body moving in a straight line and made up of a thousand subunits of equal mass. The force required to stop the motion of this body in a given period of time depends only on the total mass. It does not depend on how the subunits are distributed—whether they are packed closely together, arranged in a hollow sphere, in a cubical array, in a straight line or anything else. Only the total mass counts, and the manner in which the subunits are distributed does not change the total mass.

In rotational motion, however, it is not mass alone that counts but mass times the square of the distance from the point (or line) about which the rotation is taking place. Consider a rotating sphere, for instance, made up of a thousand subunits of equal mass. Some of the subunits are close to the axis and some are far away from the axis. Those close to the axis have a small

r, and therefore a small mr^2, while those far from the axis have a large r, and therefore a large mr^2. The body as a whole has some average mr^2, which is called the *moment of inertia*, and this is often symbolized as I. The torque required to stop the rotating sphere in a given period of time depends not upon the mass of the sphere but its moment of inertia.

The value of the moment of inertia depends on the distribution of the mass and can be changed without altering the total mass. If instead of a solid sphere we made up a hollow sphere of the same subunits, some of the subunits previously close to the axis would now be located far from the axis. On the other hand, no subunits would have been moved closer to the axis. The average r would increase and the moment of inertia (the average mr^2) would increase considerably even though the total mass had not changed. It would require a much larger torque to stop a spinning hollow sphere in a given period of time than it would to stop a solid sphere of the same mass spinning at the same angular velocity.

Thus, gyroscopes and fly-wheels in which it is desired to maintain as even an angular velocity as possible, despite torques of one sort or another, are constructed to have rims as massive as possible and interiors as light as possible. The accelerations produced by given torques are then reduced to a minimum because the moment of inertia has been raised to a maximum.

It is not surprising, considering the analogies between rotational and translational motions, that experiment shows such a thing as a *law of conservation of angular momentum*. By analogy with the law of conservation of momentum in translational motion, this additional law might be stated:

The total angular momentum of an isolated system of bodies remains constant.

But how would we define angular momentum? Ordinary translational momentum is mv, mass times velocity. For angular momentum, we must substitute moment of inertia (I) for mass, and angular velocity (ω) for translational velocity. Angular momentum, then, is equal to $I\omega$.

Again, however, the moment of inertia (the average value of mr^2) can be altered without altering the total mass, and this produces curious effects.

Suppose, for instance, that you are standing on a frictionless turntable that has been set to spinning; you are holding your arms extended, a heavy weight in each hand.

The axis of rotation is running down the center of your body from head to toe, and the mass of your extended arms is further from that axis than is the rest of you. The weights in either hand are further still. Consequently, your arms and the weights they carry, being associated with large values of r, contribute greatly to the mr^2 average and give you a much higher moment of inertia than you might ordinarily possess.

Suppose next that while spinning you lower your arms to your side. The mass content of your arms and the weights they carry is now considerably closer to the axis of rotation, and without any change in total mass, the moment of inertia is greatly decreased. If the moment of inertia (I) is decreased, the angular velocity (ω) must be correspondingly increased to keep the angular momentum ($I\omega$) constant. (In other words, if you are interested in having the product of two numbers always equal 24, then if you start with 8 times 3 and reduce the 8 to 4, you must increase the 3 to 6, to have the new numbers, 4 times 6, still equal 24.)

This, indeed, is what happens. The turntable suddenly increases its rate of spin as you bring your arms to your side. The rate decreases again promptly if you extend your arms once more.

A figure skater makes use of the same device on ice. At first, as rapid a spin as possible is produced with arms extended. The arms are then brought down, and the body spins on the point of one skate with remarkable velocity.

A body that possesses only angular momentum cannot transmit an unbalanced translational momentum to another body, for it has none to transmit. To be sure, the turning wheels of an

Conservation of angular momentum

automobile send it forward and give it translational momentum. There, however, an equal momentum is given the earth in the opposite direction. The two translational momenta add up to zero. No motorist who has ever tried to drive on ice will dismiss that fact. Once friction has decreased to the point where little or no momentum can be transmitted to the earth, the car will itself gain little or no momentum, and the wheels will spin vainly.

CHAPTER 7

Work and Energy

The Lever

Laws of conservation are popular with scientists. In the first place, a conservation law sets limits to possibilities. In considering a new phenomenon, it is convenient to be able to rule out all explanations that would involve a violation of one of the conservation laws (at least until it is found that nothing short of a violation will do). It is then easier to work with the possibilities that remain.

In addition, there is an intuitive feeling that one will not be able to get something for nothing. It therefore seems proper and orderly to suppose that the universe possesses a fixed amount of something or other (such as momentum) and that while this may be distributed among the different bodies of the universe in various ways, the total amount may neither be increased nor decreased.

Consequently, if we observe a situation in which it appears that in some respect something is obtained for nothing, a search is quickly begun for some other factor in the situation which decreases in compensation. It may prove that it is the two factors combined in some fashion that are conserved. In the case of angular momentum, for instance, the moment of inertia can be changed at will and can seemingly be made to appear out of nowhere or disappear into nowhere. The angular velocity, however, always changes in the opposite sense at once, and it is the product of the moment of inertia and the angular velocity that is conserved.

Another case of this sort arises from a consideration of the *lever*. This is any rigid object capable of turning about some fixed point called the *fulcrum*. As a practical example we might consider a wooden plank resting upon a sawhorse—the former being the lever, the latter the fulcrum.

If the fulcrum is directly under the lever's center of gravity, the lever will remain balanced, tipping neither this way nor that. Since the lever, like any other object, behaves as though all its weight were concentrated at the center of gravity, it can then be supported, as a whole, on the narrow edge of the fulcrum. If the lever is of uniform dimensions and density, the center of gravity is at the geometrical center, and it is there that the fulcrum must be placed, as in the well-known children's amusement device, the seesaw.

If a downward force is applied to any point on the lever, the force times the distance of its point of application from the fulcrum represents a torque (see page 75), and the lever takes on rotational motion in the direction of the torque.

Suppose though that a downward force is at the same time applied to the lever on the other side of the fulcrum. If the second force is equal to the first and is applied at the same distance from the fulcrum, the two torques are equal in size but not in direction. The torque on one side of the fulcrum tends to set up a clockwise rotation; the one on the other side tends to set up a counterclockwise rotation. If one torque is symbolized as τ, the other must be $-\tau$. The two torques add up to zero and the lever does not move. It remains in balance.

(On the other hand, if the force is exerted downward on one side of the fulcrum and upward on the other, then both produce a motion in the same direction: both clockwise or both counterclockwise. The torques are then both of the same sign and add up to either 2τ or -2τ. Such a doubled torque is a *couple*, and it naturally is easier to move a lever about a fulcrum by means of a couple than by means of a single torque. It is a couple we use when we wind an alarm clock or manipulate a corkscrew.)

The torques used in connection with levers are often weights that are resting on the ends of the balance, or they are on pans suspended from those ends. We can say that two equal weights will leave a lever in balance if they are placed on opposite sides of the fulcrum and at equal distances from it.

This, in fact, is the principle of the "balance." A balance

has two pans of equal weight suspended from the ends of a horizontal rod that pivots about a central fulcrum. If an object of unknown weight is placed in one pan, combinations of known weights can be put in the other till the two pans balance. We then know that the unknown weight is equal to the sum of the known weights in the other pan. (As explained on page 57, this actually serves to measure mass as well as weight.)

Because of its use in the balance, a lever subjected to equal and opposite torques is said to be in *equilibrium* (from Latin words meaning "equal weights"), and this expression has come to be applied to any system under the stress of forces that produce effects that cancel out and leave the overall condition unchanged.

For a lever to be in equilibrium it must be subjected to equal and opposite torques, and this may be true even if the forces applied are unequal. Consider a downward force (f) applied on one side of a lever at a given distance (r) from the fulcrum. The torque would be fr. Next consider a downward force twice as large ($2f$) applied to the other side of the fulcrum but at a distance only half that of the first ($-r/2$). (The distance is here given a negative sign because it is in the opposite direction from the fulcrum, as compared with the first). This second torque is $(2f)(-r/2)$, or $-fr$. The two torques are equal and opposite, and the lever remains in equilibrium.

If the forces are produced by unequal weights resting on the ends of the lever, it is easy to see that the center of gravity of the system must shift toward the end with the greater weight. To maintain equilibrium, the fulcrum must be directly under the new position of the center of gravity. When this is done, it will be found that its position is such that the product of one weight and its distance from the fulcrum will be equal to the product of the other and its distance from the fulcrum.

Thus if two children of roughly equal weight are on a seesaw, they are right to sit at the ends. If one child is markedly heavier than the other, he should sit closer to the fulcrum. The two should so distribute themselves, in fact, that their own center of gravity plus that of the seesaw remains directly above the fulcrum. (It is also possible, in the case of some seesaws, to shift the board and adjust the position of the fulcrum.)

Because of the fact that torques rather than forces must be equal in order to produce equilibrium, a lever can be put to good use. Suppose a 250-kilogram weight (equivalent to a force of about 2450 newtons) is placed 1 meter from the fulcrum. Next suppose that 10 meters from the fulcrum on the other side of the lever a man applies a downward force of 245 newtons (the equivalent of a 25-kilogram weight). The torque associated with the force (25 × 10) is equal and opposite to that of the torque produced by the weight on the other side of the lever (250 × 1). The lever is then placed in equilibrium and the heavy weight is supported by the light force. If the man applies a somewhat greater force (one that is still considerably less than that produced by the weight on the other side), the lever overbalances on his own side.

A man is not so much conscious of torque as of force (more exactly, of muscular effort). He knows that he cannot apply sufficient force directly to the 250-kilogram weight to lift it. By making use of the lever, however, he can do the job with a force one-tenth that required for direct lifting. By adjusting the differences properly, he could make do with a force one-hundredth, one-thousandth, or indeed any fraction of that required for direct lifting. The usefulness of the lever as a method of multiplying man's lift-

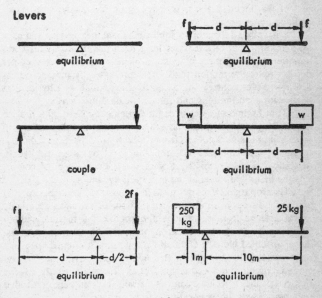

Levers

equilibrium

equilibrium

couple

equilibrium

equilibrium

equilibrium

weight × distance = weight × distance
250 × 1 = 25 × 10

ing ability is evidenced by the very word "lever," which comes from a Latin word meaning "to lift."

No doubt even primitive man had stumbled upon this "principle of the lever," but it was not until the time of the Greek mathematician Archimedes (287–212 B.C.) that the situation was analyzed scientifically. So well did Archimedes appreciate the principle of the lever and its use in unlimited multiplication of force that he said, with pardonable bombast, "Give me a place to stand on and I will move the world."

Any device that transfers a force from the point where it is applied to another point where it is used, is a *machine* (from a Latin word meaning "invention" or "device"). The lever does this, since a force applied on one side of the fulcrum can lift a weight on the other side; it does this in so uncomplicated a fashion that it cannot be further simplified. It is therefore an example of a *simple machine*. Other examples of simple machines are the inclined plane, and the wheel and axle. Some add three other simple machines to the list: the pulley, the wedge, and the screw. However, the pulley can be viewed as a sort of lever, the wedge consists of two inclined planes set back to back, and the screw is an inclined plane wound about an axis.

Virtually all the more complicated machines devised and used by mankind until recent times have been merely ingenious combinations of two or more of these simple machines. These machines depend upon the motions and forces produced by moving bodies through direct contact. As a result, that branch of physics that deals with such motions and forces is called *mechanics*.

That branch of mechanics that specifically deals with motion is called *dynamics*, while that branch that deals with motions in equilibrium is called *statics* (from a Greek word meaning "to cause to stand"). Archimedes was the first great name in the history of statics because of his work with the lever. Galileo, of course, was the first great name in the history of dynamics.

One force that does not seem to be the result of direct contact of one body upon another is gravitation. Gravitation seemingly exerts a force from a distance and produces a motion without involving direct contact between bodies. Such "action at a distance" troubled Newton and many physicists after him. Expedients were worked out to explain this away, and gravitation was included among the mechanical forces. Thus, the study of the motions of the heavenly bodies that result from and are controlled by gravitational forces is called *celestial mechanics*.

Multiplying Force

A machine not only transfers a force, it can often be used to multiply that force, as in the example of the lever described above. Yet this multiplication of force should be approached with suspicion. How can one newton of force do the work of ten newtons just by transmitting it through a rigid bar? Such generosity on the part of the universe is too much to expect, as I pointed out at the beginning of the chapter. Something else must be lost to make up for it.

If we consider the lever lifting the 250-kilogram weight by use of a force equivalent to only 25 kilograms of weight, we can see in the accompanying diagram that we have two similar triangles. The sides and altitude of one are to the corresponding sides and altitude of the other as the distance of the weight from the fulcrum is to the distance of the applied force from the fulcrum.

In other words, if we apply a force at a point ten times as far from the fulcrum as the weight is, then to lift the weight a given distance, we must push down through a distance ten times as great. There is the answer! In lifting a weight by means of a lever, we may adjust distances from the fulcrum in such a way as to make use of a fraction of the force that would be required without the lever, but we must then apply that fractional force through a correspondingly greater distance. The product of the force multiplied by the distance remains the same at either end of the lever.

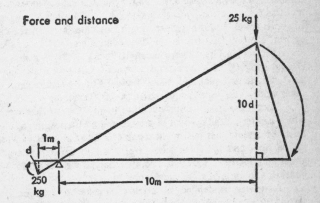

Force and distance

25 kg

10 d

1m

d

250 kg

10m

This turns out to be true of any machine that seems to multiply a force. The smaller force performs the task that would require a larger force without the machine, but always at the cost of having to be exerted through a correspondingly longer distance. The product of force and distance in the direction of the force is called *work* and is usually symbolized as *w*, so that:

$$w = fd \qquad \text{(Equation 7-1)}$$

In a sense, work is an unfortunate term to use in this connection. Anyone will agree that lifting a weight through a distance is work, but in the common use of the term matters are not confined to this alone. In the common language, work is a term applied to the product of any form of exertion. To sit quietly in my chair for half an hour and think of what I am going to say next in this book may strike me as being hard work, but it involves no action of a force through a distance and is not work to a physicist. Again, to stand in one place and hold a heavy suitcase seems hard work, but since the suitcase doesn't move, no work is being done on it. If one walks along with the suitcase, there is still no work being done on it, for although it is moving (horizontally), it is not moving in the direction of the (vertical) force that keeps it from falling.

Nevertheless, the term *work*, signifying a force multiplied by the distance through which a body moves in the direction of the force, is ineradicably established and must be accepted.

The units of work are those of force multiplied by those of distance. In the mks system, the unit of work is the newton-meter, and this is named the *joule* (pronounced "jool") after an English physicist whom I will have occasion to mention later. In the cgs system, the unit of work is the dyne-centimeter, which is called the *erg* (from a Greek word meaning "work"). Since a newton is equal to 100,000 dynes and a meter to 100 centimeters, a newton-meter is equal to 100,000 times 100 dyne-centimeters. In other words, one joule is equal to 10,000,000 ergs.

Since force is a vector quantity, it might seem that work, which is after all the product of a force and a distance, might also be a vector; and that one might speak of a given amount of work to the right and the same amount of work to the left as being equal and opposite. This is not so, however, as we will find if we consider the units of work once more.

A newton is defined as a kilogram-meter per second per second, or $kg\text{-}m/sec^2$. If a joule is a newton-meter, then it is also a kilogram-meter-meter per second per second, or a kg-m²/sec². This last can be written kg-(m/sec)². But m/sec (meters per second) is a unit of velocity, and this means that the unit of work is equal to the unit of mass times the square of the unit of velocity, or $w = mv^2$.

It is true that velocity is a vector quantity, therefore one might speak of $-v$ and $+v$, but the unit of work involves the square of the velocity. The square of a positive number $(+v)$ $(+v)$ and the square of a negative number $(-v)(-v)$ are both positive $(+v^2)$, as we know from elementary algebra. Consequently, the square of the velocity involves no differences in signs, and a unit that includes the square of the velocity is not a vector unit (unless it contains vector units other than velocity, of course).

We conclude then that work is a scalar quantity.

Returning to the lever, we see that the work involved in raising a boulder with a lever is the same as that involved in raising a boulder without a lever, but that the distribution of work between force and distance differs. The same is true where an inclined plane is the device used.

Let us say it is necessary to raise a 50-kilogram barrel through a height of two meters onto the back of a truck. Since a kilogram of weight exerts a downward force of 9.8 newtons, a total upward force of 490 newtons is required to lift the barrel. To exert 490 newtons of force through a distance of two meters in the direction of the force is to do 980 joules of work.

Suppose instead that we lay a sloping plank from the ground to the truck so that the plank makes an angle of 30° with the ground. Under those conditions, the length of the plank from ground to truck is just twice the vertical height from ground to truck, or four meters. The force required to roll the barrel up the plank is 295 newtons, just half the force required for direct lift-

Inclined plane

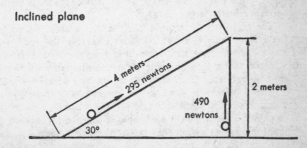

4 meters
295 newtons
490 newtons
2 meters
30°

ing. That half-force is exerted through double-distance, however, and 980 joules of work is still done.

The gentler the slope of the inclined plane, the smaller the force required to move the barrel and the longer the distance through which it must be moved. The inclined plane dilutes force —as it diluted velocity for Galileo—by diluting gravitational force (see page 10). Neither the inclined plane, nor the lever, nor any machine, dilutes work. If we stick to work, rather than force, we never get something for nothing.

But if we gain nothing on work, why bother? The answer is that even if we gain nothing directly, we may gain by altering the distribution between force and distance. If it is a question of lifting, by our unaided effort, 250 kilograms two meters directly upward, we must give up. We cannot lift it a meter, a centimeter, or anything at all; we cannot budge it. To move the equivalent of 50 kilograms through ten meters is possible, however, especially if we work slowly; in this way we can do the same work (50×10) that would have been impossible under the previous conditions (250×2). To lift the equivalent of five kilograms through 100 meters may be tedious, but it is quite easy.

Again, if we were asked to shinny up a rope suspended from the roof of a five-story building, we might well decide it to be beyond our capacity unless we were in excellent physical shape. However, a quite ordinary man can lift his weight to a fifth-story roof, if he goes up by way of a ramp, which is an inclined plane that enables him to use less force to lift his body—at the expense of moving it through a longer distance.

It is sometimes convenient to do the opposite: expend extra force in order to gain distance. Thus, a great deal of force is exerted upon the pedals of a bicycle. This is transmitted to a point on the rear wheel near the hub. The spokes of the wheel then act as levers (with the hub the fulcrum), so a much smaller force is applied to the rim of the wheel which, however, moves through a correspondingly larger distance.

The bicycle is therefore a machine that enables the body to convert force into distance (without changing the total work done) more efficiently than it could without the bicycle. It is for this reason that a man on a bicycle can easily outrace a running man, although both are using their leg muscles with equal effort.

The definition of work as the product of a force and the distance through which it acts, says nothing about the time it takes to act. Men usually find it preferable to accomplish a particular amount of work in a short time rather than in a long time and are therefore interested in the rate at which work is done. This rate is spoken of as *power*. The units of power are joules/sec in the mks system and ergs/sec in the cgs system.

A very common unit of power which fits into neither system was originated by the Scottish engineer James Watt (1736–1819). He had improved the steam engine and made it practical toward the end of the eighteenth century, and he was anxious to know how its rate of work in pumping water out of coal mines compared with the rate of work of the horses previously used to operate the pumps. In order to define a *horsepower* he tested horses to see how much weight they could lift through what distance and in what time. He concluded that a strong horse could lift 150 pounds through a height of 220 feet in one minute, so one horsepower was equal to $150 \times 200/1$, or 33,000 foot-pounds/minute.

This inconvenient unit is equal to 745.2 joules/sec, or 7,452,-000,000 ergs/sec. A joule/sec is defined as a *watt* in James Watt's honor, and so we can also say that one horsepower is equal to 745.2 watts. The watt, however, is most commonly used in electrical measurements. In mechanical engineering (at least in Great Britain and the United States) it is still horsepower all the way. The power of our automobile engines, for instance, is routinely given in horsepower.

Mechanical Energy

It is neat and pleasant to see that the work put into one end of a lever is equal to the work coming out of the other end, and we might fairly suspect that there was such a thing as "conservation of work."

Unfortunately, such a possible conservation law runs into a snag almost at once. After all, where did the work come from that was put into the lever? If one end of the lever was manipulated by a human being who was using the lever to lift a weight, the work came from that done by the moving human arm.

And where does the work of the moving arm come from? A man sitting quietly can suddenly move his arm and do work where no work had previously seemed to exist. This runs counter to the notion of conservation in which the phenomenon being conserved can be neither created nor destroyed.

If one is anxious to set up a conservation law involving work, therefore, one might suppose that work, or something equivalent to work, could be stored in the human body (and perhaps in other objects) and that this work-store could be called upon at need and converted into visible, palpable work.

At first blush such a work-store might have seemed to be particularly associated with life, since living things seemed filled with this capacity to do work, whereas dead things, for the most part, lay quiescent and did not work. The German philosopher and scientist Gottfried Wilhelm Leibnitz (1646–1716), who was the first to get a clear notion of work in the physicist's sense, chose to call this work-store *vis viva* (Latin for "living force").

However, it is clearly wrong to suppose that work is stored only in living things; as a matter of fact, the wind can drive ships and running water can turn millstones, and in both cases force is being exerted through a distance. Work, then, was obviously stored in inanimate objects as well as in animate ones. In 1807, the English physician Thomas Young (1773–1829) proposed the term *energy* for this work-store. This is from Greek words meaning "work-within" and is a purely neutral term that can apply to any object, living or dead.

This term gradually became popular and is now applied to any phenomenon capable of conversion into work. There are many varieties of such phenomena and therefore many forms of energy.

The first form of energy to be clearly recognized as such, perhaps, was that of motion itself. Work involved motion (since an object had to be moved through a distance), so it was not surprising that motion could do work. It was moving air, or wind, that drove a ship, and not still air; moving water that could turn a millstone, and not still water. It was not air or water that contained energy then, but the motion of the air or water. In fact, anything that moved contained energy, for if the moving object, whatever it was, collided with another, it could transfer its momentum to that second object and set its mass into motion; it would thus be doing work upon it, for a mass would have moved through a distance under the urging of a force.

The energy associated with motion is called *kinetic energy*, a term introduced by the English physicist Lord Kelvin (1824–1907) in 1856. The word "kinetic" is from a Greek word meaning "motion."

Exactly how much kinetic energy is contained in a body moving at a certain velocity, v? To determine this, let us assume that in the end we are going to discover that there exists a conservation law for work in all its forms—stored and otherwise. In that case, we can be reasonably confident that if we find out how much work it takes to get a body moving at a certain velocity, v, then that automatically will be the amount of work it can do on some other object through its motion at that velocity. In short, that would be its kinetic energy.

To get a body moving in the first place takes a force, and that force, by Newton's second law, is equal to the mass of the moving body multiplied by its acceleration: $f = ma$. The body will travel for a certain distance, d, before the acceleration brings it up to the velocity, v, which we are inquiring into. The work done on the body to get it to that velocity is the force multiplied by the distance. Expressing the force as ma we have:

$$w = mad \qquad \text{(Equation 7-2)}$$

Now much earlier in the book, in discussing Galileo's experiments with falling bodies, we showed that $v = at$—that velocity, in other words, is the product of acceleration and time. This is easily rearranged to: $t = v/a$. We also pointed out in discussing Galileo's experiments that where there is uniform acceleration, $d = \frac{1}{2}at^2$, where d is the distance covered by the moving body. If, in place of t in the relationship just given, the quantity v/a is substituted, we have:

$$d = \frac{1}{2} a \left(\frac{v}{a}\right)^2 = \frac{1}{2} \frac{v^2}{a} \qquad \text{(Equation 7-3)}$$

Let us now substitute this value for d in Equation 7-2, which becomes:

$$w = \frac{1}{2} \frac{mav^2}{a} = \frac{1}{2} mv^2 \qquad \text{(Equation 7-4)}$$

This is the work that must be done upon a body of mass m to get it to move at a velocity v, and it is therefore the kinetic energy contained by the body of that mass and with that velocity. If we symbolize kinetic energy as e_k, we can write:

$$e_k = \frac{1}{2} mv^2 \qquad \text{(Equation 7-5)}$$

I have already pointed out that work has the units of mass multiplied by those of velocity squared and, as is clear from Equation 7-5, so has kinetic energy. Therefore, kinetic energy can be measured in joules or ergs, as can work. Indeed, all forms of energy can be measured in these units.

We might now imagine that we can set up a conservation

law in which kinetic energy can be converted into work and vice versa, but in which the sum of kinetic energy and work in any isolated system must remain constant. Such a conservation law will not, however, hold water, as can easily be demonstrated.

An object thrown up into the air has a certain velocity and therefore a certain kinetic energy as it leaves the hand (or the catapult or the cannon). As it climbs upward, its velocity decreases because of the acceleration imposed upon it by the earth's gravitational field. Kinetic energy is therefore constantly disappearing and, eventually, when the ball reaches maximum height and comes to a halt, its kinetic energy is zero and has therefore entirely disappeared.

One might suppose that the kinetic energy has disappeared because work has been done on the atmosphere, and that therefore kinetic energy has been converted into work. However, this is not an adequate explanation of events, for the same thing would happen in a vacuum.

One might next suppose that the kinetic energy had disappeared completely and beyond redemption, without the appearance of work, and that no conservation law involving work and energy was therefore possible. However, after an object has reached maximum height and its kinetic velocity has been reduced to zero, it begins to fall again, still under the acceleration of gravitational force. It falls faster and faster, gaining more and more kinetic energy, and when it hits the ground (neglecting air resistance) it possesses all the kinetic energy with which it started.

Rather than lose our chance at a conservation law, it seems reasonable to assume that energy is not truly lost as an object rises upward, but that it is merely stored in some form other than kinetic energy. Work must be done on an object to lift it to a particular height against the pull of gravity, even if once it has reached that height it is not moving. This work must be stored in the form of an energy that it contains by virtue of its position with respect to the gravitational field.

Kinetic energy is thus little by little converted into "energy of position" as the object rises. At maximum height, all the kinetic energy has become energy of position. As the object falls once more, the energy of position is converted back into kinetic energy. Since the energy of position has the potentiality of kinetic energy, the Scottish engineer William J. M. Rankine (1820–1872) suggested, in 1853, that it be termed *potential energy*, and this suggestion was eventually adopted.

To lift a body a certain distance (d) upward, a force equal to its weight must be exerted through that distance. The force exerted by a weight is equal to mg, where m is mass and g the acceleration due to gravity (see Equation 5–1 on page 54). If we let potential energy be symbolized as e_p, then, we have:

$$e_p = mgd \qquad \text{(Equation 7–6)}$$

If all the kinetic energy of a body is converted into potential energy, then the original e_k is converted into an equivalent e_p, or combining Equations 7–5 and 7–6:

$$\frac{1}{2} mv^2 = mgd$$

or simplifying, and assuming g to be constant.

$$v^2 = 2gd = 19.6d \qquad \text{(Equation 7–7)}$$

From this relationship one can calculate (neglecting air resistance) the height to which an object will rise if the initial velocity with which it is propelled upward is known. The same relationship can be obtained from the equations arising out of Galileo's experiments with falling objects.

Kinetic energy and potential energy are the types of energy made use of by machines built up out of levers, inclined planes and wheels, and the two forms may therefore be lumped together as *mechanical energy*. As long ago as the time of Leibnitz it was recognized that there was a sort of "conservation of mechanical energy," and that (if such extraneous factors as friction and air resistance were neglected) mechanical energy could be visualized as bouncing back and forth between the kinetic form and the potential form, or between either and work, but not (taken in all three forms) as appearing from nowhere or disappearing into nowhere.

The Conservation of Energy

Unfortunately, the "law of conservaton of mechanical energy," however neat it might seem under certain limited circumstances, has its imperfections, and these at once throw it out of court as a true conservation law.

An object hurled into the air with a certain kinetic energy, returns to the ground without quite the original kinetic energy. A small quantity has been lost through air resistance. Again, if an elastic object is dropped from a given height, it should (if mechanical energy is to be truly conserved) bounce and return to exactly its original height. This it does not do. It always returns to somewhat less than the original height, and if allowed to drop again and bounce and drop again and bounce, it will reach lower and lower heights until it no longer bounces at all. Here it is not only the air resistance that interferes but also the imperfect elasticity of the body itself. Indeed, if a lump of soft clay is dropped, its potential energy is converted to kinetic energy, but at the moment it strikes the ground with a nonbouncing splat that kinetic energy is gone—and without any re-formation of potential energy. To all appearances, mechanical energy disappears in these cases.

One might argue that these losses of mechanical energy are due to imperfections in the environment. If only a frictionless system were imagined in a perfect vacuum, if all objects were completely elastic, then mechanical energy would be conserved.

However, such an argument is quite useless, for in a true conservation law the imperfections of the real world do not affect the law's validity. Momentum is conserved, for instance, regardless of friction, air resistance, imperfect elasticity or any other departure from the ideal.

If we still want to seek a conservation law that will involve work, we must make up our minds that for every loss of mechanical energy there must be a gain of something else. That something else is not difficult to find. Friction, one of the most prominent imperfections of the environment, will give rise to heat, and if the friction is considerable, the heat developed is likewise considerable. (The temperature of a match-head can be brought to the ignition point in a second by rubbing it against a rough surface.)

Conversely, heat is quite capable of being turned into mechanical energy. The heat of the sun raises countless tons of water vapor kilometers high into the air, so that all the mechanical energy of falling water (where as rain, cataracts or quietly flowing rivers) must stem from the sun's heat. Futhermore, the eighteenth century saw man deliberately convert heat into mechanical energy by means of a device destined to reshape the world. Heat was used to change water into steam in a confined chamber, and this steam was then used to turn wheels and drive pistons. (Such a device is, of course, a steam engine.)

It seemed clear, therefore, that one must add the phenomenon of heat to that of work, kinetic energy and potential energy, in working out a true conservation law. Heat, in short, would have to be considered another form of energy.

But if that is so, then any other phenomenon that could give rise to heat would also have to be considered a form of energy. An electric current can heat a wire and a magnet can give rise to an electric current, so both electricity and magnetism are forms of energy. Light and sound are also forms of energy, and so on.

If the conservation law is to encompass work and all forms of energy (not mechanical energy alone), then it had to be shown that one form of energy could be converted into another quantitatively. In other words, in such energy-conversions all energy must be accounted for; no energy must be completely lost in the process, no energy created.

This point was tested thoroughly over a period of years in the 1840's by an English brewer named James Prescott Joule (1818-1889), whose hobby was physics. He measured the heat produced by an electric current, that produced by the friction of water against glass, by the kinetic energy of turning paddle wheels in water, by the work involved in compressing gas, and so on. In doing so, he found that a fixed amount of one kind of energy was converted into a fixed amount of another kind of energy, and that if energy in all its varieties was considered, no energy was either lost or created. It is in his honor that the unit of work and energy in the mks system is named the "joule."

In a more restricted sense, one can consider that Joule proved that a certain amount of work always produced a certain amount of heat. Now the common British unit of work is the "foot-pound" —that is, the work required to raise one pound of mass through a height of one foot against the pull of gravity. The common British unit of heat is the "British thermal unit" (commonly abbreviated "Btu") which is the amount of heat required to raise the temperature of one pound of water by 1° Fahrenheit. Joule and his successors determined that 778 foot-pounds are equivalent to 1 Btu, and this is called the *mechanical equivalent of heat*.

It is preferable to express this mechanical equivalent of heat in the metric system of units. A foot-pound is equal to 1.356 joules, so 778 foot-pounds equal 1055 joules. Furthermore, the most common unit of heat in physics is the *calorie*, which is the amount of heat required to raise the temperature of one gram of water by 1° Centigrade.* One Btu is equal to 252 calories. Therefore, Joule's mechanical equivalent of heat can be expressed as 1055 joules equal 252 calories, or 4.18 joules = 1 calorie.

Once this much was clear, it was a natural move to suppose that the law of conservation of mechanical energy should be con-

---

* There will be more to say about Fahrenheit degrees, Centigrade degrees, calories, and other items of the sort later in the book; see chapters 13 and 14.

verted into a *law of conservation of energy*, in the broadest sense of the word—including under "energy," work, mechanical energy, heat, and everything else that could be converted into heat. Joule saw this, and even before his experiments were far advanced, a German physicist named Julius Robert von Mayer (1814–1878) maintained it to be true. However, the law was first explicitly stated in form clear enough and emphatic enough to win acceptance by the scientific community in 1847 by the German physicist and biologist Hermann von Helmholtz (1821–1894), and it is he who is generally considered the discoverer of the law.

The law of conservation of energy is probably the most fundamental of all the generalizations made by scientists and the one they would be most reluctant to discard. As far as we can tell it holds through all the departures of the real universe from the ideal models set up by scientists; it holds for living systems as well as nonliving ones; and for the tiny world of the subatomic realm as well as for the cosmic world of the galaxies. At least twice in the last century phenomena were discovered which seemed to violate the law, but both times physicists were able to save matter by broadening the interpretation of energy. In 1905, Albert Einstein showed that mass itself was a form of energy; and in 1931, the Austrian physicist Wolfgang Pauli (1900–1958) advanced the concept of a new kind of subatomic particle, the neutrino, to account for apparent departures from the law of conservation of energy.

Nor was this merely a matter of saving appearances or of patching up a law that was springing leaks. Each broadening of the concept of conservation of energy fit neatly into the expanding structure of twentieth-century science and helped explain a host of phenomena; it also helped predict (accurately) another host of phenomena that could not have been explained or predicted otherwise. The nuclear bomb, for instance, is a phenomenon that can only be explained by the Einsteinian concept that mass is a form of energy.

Vibration

Simple Harmonic Motion

The law of conservation of energy serves to throw light on a type of motion that we have not yet considered.

So far, the motions that have been discussed, whether translational or rotational, have progressed (unless disturbed) in one direction continuously. It is, however, also possible for motion to progress alternately, first in one direction, then in another, changing direction sometimes after long intervals and sometimes after short intervals—even very short intervals. Such an alternate movement in opposite directions is called a *vibration* or *vibratory motion* (from a Latin word meaning "to shake").

This type of motion is very common, and we are constantly aware of the swaying or trembling of branches and leaves in the wind, for instance; or of the rapid trembling of machinery in operation, such as that of an automobile with its motor idling; even of the chattering of our teeth or the shaking of our hand under conditions of cold or of nervous tension.

The form of vibration that first came under scientific scrutiny was that of a taut string when plucked. Such strings were used in musical instruments known even to the ancients; the plucked strings give rise to musical sounds for reasons involving the vibratory motions lent by the vibrating strings to the air itself (see the chapters on sound beginning on page 148). The first to study such vibrations was the ancient Greek mathematician and philosopher Pythagoras of Samos (sixth century B.C.). His interest lay entirely in the relationship of these vibrations to music, and as a result, vibratory motion is frequently called *harmonic motion*.

Most vibratory motion is of a complicated nature and does not readily lend itself to easy mathematical analysis. The particular type exemplified by the taut, vibrating string is, however, an exception. It can be analyzed with comparative ease and is therefore called *simple harmonic motion* (sometimes abbreviated SHM).

In simple harmonic motion, it has been found that Hooke's law (see page 50) holds at every stage of the movement. If we pull a taut string out of its original equilibrium position, the amount of the displacement from that equilibrium position is proportional to the force tending to restore it to the equilibrium position.

If the string is released after being pulled to the right, let us say, the restoring force accelerates it in the direction of the equilibrium position. In other words, the string snaps back to equilibrium, moving faster and faster as it does so.

As it approaches the equilibrium position, its displacement from that position becomes continually less, and the restoring force becomes continually less in proportion. As the restoring force decreases so, naturally, does the acceleration it imparts; therefore, although the string moves more and more rapidly as it approaches the equilibrium position, the rate of gain of velocity becomes less and less. Finally, when it has reached the equilibrium position the restoring force has become zero and so has acceleration. The string can gain no more velocity and its rate of motion is at a maximum.

But although it is no longer gaining velocity it *is* moving rapidly, and it cannot remain at equilibrium position, but must move past it. Only a force can stop it once it is moving (Newton's first law), and at equilibrium position there is no force to do so. As it goes past the point of equilibrium to the left, however, it is displaced once more, and a restoring force comes into being again; this force produces an acceleration that serves to diminish its velocity of movement (which is now in the direction opposed to the force). As the string continues to move leftward, the displacement and the restoring force continue to increase, and the velocity diminishes at a faster and faster rate until it reaches zero. The string is now motionless at a point of maximum leftward displacement that is equal to the original extent of the rightward displacement.

Under the influence of the restoring force, the string moves to the right again, passes through the equilibrium position, and out to the original maximum rightward displacements. Then it goes back to the left, then back to the right, and so on.

If there were no air resistance and no friction at the points where the string is held taut, the maximum displacements to left and right would always be the same and the vibration would continue indefinitely. As it is, the vibrations do not, after all, quite reach the maximum, but with each rightward (or leftward) motion attain a point of displacement not quite equal to the point reached at the previous motion in that direction. The vibrations are "damped" and slowly die out.

In all cases of simple harmonic motion, the crucial fact is that velocity changes smoothly at all times, never abruptly. Suppose one imagines a falling body passing through the surface of the earth and the solid substance of the planet. The gravitational force upon it would grow continually less as more and more of the substance of the planet lay above the falling body and less and less below it. The body would accelerate as it fell, but by a smaller and smaller amount. By the time it reached the center of the earth, there would be no force upon it at all (at that point), and its velocity would be at a maximum. It would then pass beyond the earth's center and begin to emerge through the opposite portion of the planet, its velocity decreasing as the gravitational force grew larger and larger, until it emerged from the surface at the antipodes and rose as high above it as it had been (on the other side) in the beginning. It would then repeat this movement, returning to its original position, then to the opposite position, and so on. This, too, would be an example of simple harmonic motion.

In actual fact, however, the falling body is interrupted by the surface of the earth, and its velocity is abruptly changed at the moment of contact with that surface. The resultant series of bounces, while an example of a vibratory or harmonic motion, is not a simple harmonic motion.

The Period of Vibration

A particular point of interest in any vibratory motion is the time it takes to move from the extreme point on one side to the extreme point on the other and back. The time taken to complete this motion (or any particular motion, for that matter) is the *period* of that motion.*

Whenever a motion goes through a series of repetitive submotions, each with a period of its own, the motion is said to be a *periodic motion*, particularly when the individual periods are equal. Motion about a circle or any closed curve can be viewed as made up of successive returns to an original point with each single movement about the curve; it is hence a series of repetitive submotions and may be a periodic motion. A vibration also represents a series of returns to an original point, though by way of a forward-and-back motion rather than by motion in a closed curve, and a vibration can also be a periodic motion.

To determine the period of a vibrating object, even when it is vibrating in accordance with the laws governing simple harmonic motion, is rather complicated if the vibration is dealt with directly. In such a vibration, neither velocity nor acceleration is constant, but both are changing with position at every instant. One therefore searches for a way of representing a vibration by means of some sort of motion involving a constant acceleration.

This can be done by switching from vibration to another form of periodic motion—that of motion in a circle. An object can be pictured as moving in a circle under constant inward acceleration, and hence as moving along the circumference of the circle at a constant speed.

If the circle in question has a radius of length a, then its circumference is $2\pi a$. If the point is moving at a speed v, then the time, t, it takes to make a complete revolution (the period of the circular motion) is:

$$t = \frac{2\pi a}{v} \qquad \text{(Equation 8–1)}$$

Now if we imagine the circle casting a shadow edge-on upon the wall, its shadow will be that of a straight line. The point moving about the circle will seem in the shadow to be moving back and forth on the straight line. As the point moves once about the circle, the point on the shadow will seem to move once back and forth upon the straight line. The period of the motion about the circle (Equation 8–1) will also be the period of the shadow-vibration.

---

* The word "period" comes from Greek words meaning "round path" or "circle," because the first motion to interest mankind from the standpoint of the time it took was, of course, the apparent circular motion of the sun across the sky from one sunrise to the next.

At either extremity of the shadow-line, the point will seem to be moving very slowly, for its motion on the circle will be more or less at right angles to the shadow-line, and there will be very little sideways motion. (And only sideways motion will show up on the shadow.) As the point travels into intermediate parts of the circle, more and more of its motion is sideway and less and less toward or away from the line, so the point on the shadow-line seems to move faster and faster the further it is from the extremity. At the very center, the point on the circle is moving quite parallel to the line and all its motion is sideway. At the center of the shadow line, therefore, the point seems to be moving fastest. The motion of the point on the shadow-line seems to resemble that of a body in simple harmonic motion and, indeed, the motion can be shown to *be* that of a body in simple harmonic motion. Consequently, Formula 8–1 represents the period (t) of a simple harmonic motion.

Equation 8–1 still represents a difficulty for it involves v, a velocity, and while the point travels about the circle at uniform speed, it moves on the shadow-line with a constantly changing one. We must, therefore, find something to substitute for v, if we can.

In any simple harmonic motion, the maximum velocity comes at the midpoint, between the two extremes. A body undergoing such motion is then at equilibrium position, where it would

Periodic motions

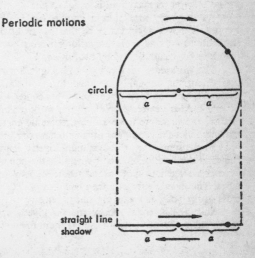

circle

a   *a*

straight line
shadow

a   *a*

remain if it were at rest. It has no energy of position at that point, and whatever energy it has is all energy of movement, or (see page 94) kinetic energy. As the object moves away from its equilibrium position, it loses velocity and therefore loses kinetic energy. However, it moves into a position it would not take up but for the kinetic energy, and so it gains energy of position, or (see page 96) potential energy. At the extreme position, the body comes to a momentary halt, and all its energy is in the form of potential energy. A body in simple harmonic motion demonstrates a periodic shift from kinetic energy to potential energy and back again, and (barring the damping effect of friction and air resistance) is an excellent example of a conservation of mechanical energy.

Now I have already said that, by Hooke's law, the restoring force on a body undergoing simple harmonic motion is proportional to its displacement from equilibrium position. That is $f = kd$, where f is the restoring force and d is the displacement. The restoring force is least at the position of equilibrium (which is at the center of our straight-line shadow). There is no displacement at that point and the force is equal to 0. The restoring force is at a maximum at the point of maximum displacement, which is, of course, at the extremity of the straight-line shadow. That extremity is a distance of a (the radius of the circle that casts the straight-line shadow) from the center, or equilibrium position, so the force at its maximum is equal to ka.

As the body moves from equilibrium position to the extremity, it moves against a force that begins at 0 and increases smoothly to ka, and the average force against which it moves is therefore ka plus 0 divided by two, or $ka/2$.

The work done on the body to bring it from its equilibrium position to the extremity is equal to force times the distance through which the force is exerted. This comes to $ka/2$ times a, or $ka^2/2$. At the extreme point, all this work is stored as potential energy, and therefore the maximum potential energy of the body moving under the conditions of simple harmonic motion is $ka^2/2$.

At the same time, the maximum kinetic energy of the body comes at the equilibrium point where all the potential energy has been converted into motion and where velocity reaches a maximum. The kinetic energy is then equal to $mv^2/2$, where m is the mass of the body and v its maximum velocity.

Since the potential energy and the kinetic energy are interconverted constantly during simple harmonic motion, without significant loss, the maximum potential energy and the maximum kinetic energy must be equal, so:

$$mv^2/2 = ka^2/2 \qquad \text{(Equation 8-2)}$$

We can easily rearrange this equation to:

$$\frac{a}{v} = \sqrt{\frac{m}{k}} \qquad \text{(Equation 8-3)}$$

We can substitute $\sqrt{m/k}$ for a/v in Equation 8-1 and we have:

$$t = 2\pi \sqrt{\frac{m}{k}} \qquad \text{(Equation 8-4)}$$

This is an astonishing result, for the period of simple harmonic motion turns out to depend only on the mass of the moving body and on the proportionality constant between stress and strain. Both can easily be determined for a particular body, and the period can then be calculated at once.

The period, it should be noted, does *not* depend on the velocity of the body moving in simple harmonic motion, nor on the amount by which it is displaced from equilibrium position, since both v and a have disappeared from Equation 8-4. This means that if a string is pulled out from its equilibrium position by a certain amount, it will attain a certain maximum velocity at mid-point of its swings and will have a certain period of vibration. If it is pulled out a greater distance, or a lesser distance, it will gain a greater maximum velocity, or a lesser one, respectively; in either case the change in velocity will be just enough to make up for the change in distance of displacement, so the period will remain the same.

This constant period of vibration offers mankind a great boon; it is a means of measuring time quite accurately by counting vibrations, even damped vibrations.

In theory, any periodic motion makes this possible. The first periodic motion to serve mankind as a timepiece was the earth itself, for each turn of the planet on its axis marks off one day and night and each turn of the planet about the sun marks off one cycle of seasons. Unfortunately, the earth's movements do not offer a good means of measuring times of less than a day.

During ancient times, mankind made use of nonperiodic motions broken up into (as was hoped) equal parts. These included the motion of a shadow along a background, the movement of sand through a narrow orifice, the dripping of water through an orifice, the shrinking of a burning candle, and so on. All that was obtained in this fashion were rather poor approximations of equal times; not until the mid-seventeenth century was it possible to tell time to closer than an hour or so, or to measure units of time less than an hour with any reasonable accuracy.

It was not until periodic motions with short periods of vibration were put to use that modern time-telling devices became possible—and with them, to a very large extent, modern science.

The Pendulum

Galileo himself suffered greatly from the inability to measure short intervals of time accurately. (He made use of his pulse on occasions, and though this was a periodic phenomenon, it was not, unfortunately, a very steady one.) Nevertheless, although he was himself not to benefit directly from it, he was the first to discover a periodic motion that was eventually to be put to use for the purpose of time-telling.

In 1583, Galileo was a teen-age medical student at the University of Pisa and one day went to the cathedral to pray. Even his devotion to prayers (and Galileo was always a pious man) could not keep his agile mind from working. He could not help but notice the chandelier swaying in the draft. At times, thanks to the vagary of the wind, it swayed in large arcs, at times in smaller ones, but it seemed to Galileo that the period of swing was always the same, regardless of the length of the arc. He interrupted his prayers and checked this conjecture by timing the swing against his pulse.

Back at his quarters, Galileo went on to set up small experimental "chandeliers" by suspending heavy weights ("bobs") from strings attached to the ceiling and letting them swing to and fro. (Such suspended weights are called *pendulums* from a Latin word meaning "hanging" or "swinging"). Galileo was able to show that the period of swing did not depend on the weight of the bobs, but only upon the square root of the length of the string. In other words, a pendulum with a string four feet long would have a period twice as long as one with a string one foot long.

Consider the pendulum, now. If the bob is suspended vertically from its support, it will remain motionless. That is its equilibrium position. If, however, the bob is pulled to one side, the pull of the string forces it to move in the arc of a circle so that it is raised to a higher level. If it is now released, the pull of gravity will cause it to move downward with an accelerating velocity, back along the arc of the circle to the bottommost position.

The gravitational force that brings about this fall is only that part not balanced by the upward pull of the string. As the bob drops, the string becomes more and more nearly vertical and balances more and more of the gravitational force. The unbalanced gravitational force constantly decreases as the bob drops, and the acceleration to which the bob is subjected also decreases. When the bob is at the bottom of the arc, the pendulum is perfectly vertical and the string balances all the gravitational pull. There is no unbalanced gravitational pull at that point and no acceleration. The bob is moving at maximum velocity.

Because of inertia, the bob passes through the point of equilibrium and begins to mount the arc in the other direction. Now there is an unbalanced gravitational force that slows its motion. The higher it climbs the greater the unbalanced gravitational force and the more quickly is the motion of the bob slowed down. Eventually its motion is slowed to zero and it reaches a maximum displacement. Down it comes again, through the equilibrium point, to a maximum displacement on the other side, and so on.

This is very much like the discription of simple harmonic motion (see page 102), except that where the plucking of a string involves motion back and forth in a straight line, that of the pendulum involves motion back and forth along a circular arc. This in itself would not seem to be an essential difference, since there seems no reason why there should not be vibratory rotational motion as well as vibratory translational motion; indeed there are cases of both varieties of simple harmonic motion.

Pendulum

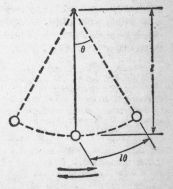

But is the pendulum truly one of them? In all cases of simple harmonic motion such as the vibration of a string, the twisting and untwisting of a wire, the up and down movement of a stretched string, and the opening and closing of a spirally coiled spring, the restoring force lies within the material itself; it is the product of its elasticity. In the case of the pendulum, the restoring force lies outside the system in the form of an unbalanced gravitational pull. This may well introduce a fundamental difference. To check on whether the pendulum swings according to simple harmonic motion, we must see whether the restoring force of gravity is indeed proportional to the amount of displacement, which is what would be required if Hooke's law (characteristic of simple harmonic motion) is to hold.

Let us begin with the displacement. This is the length of the circular arc through which the pendulum has moved in reaching a particular position. The length of this arc depends both on the length (l) of the string and on the size of the angle (θ)* through which it moves. The displacement (D) is, in fact, equal to the length of the string times the angle through which the weight moves:

$$D = l\theta \qquad \text{(Equation 8–5)}$$

Now what about the restoring force? That depends upon the force of gravity, of course. The full pull of gravity, directed downward, would be equal to mg (see page 54), where m is the mass of the bob and g is the gravitational acceleration.† However, the bob is not being pulled directly downward, but to one side. It moves as though it were sliding down an inclined plane that changes its slope at every point.

The situation is similar to that on page 18, where we were also involved with inclined planes. Imagine the bob of a pendulum at a certain point of its movement, where the suspending string makes an angle θ with the vertical. At that point, the bob is acting as though it were sliding down an inclined plane that made a

---

* The Greek letter, "theta" (θ) is often used to represent angles.

† Actually, the string also has a mass, however light that may be, so there is mass distributed all along the line of the pendulum from the bob up to the very support. At each point in the string there is a little bit of mass suspended by a different length of string. This is also true of the bob itself, different portions of which are different distances from the point of support. Ideally, a pendulum should consist of a massive bob with zero volume attached by a weightless string to the point of support. Such a device is an ideal or *simple pendulum*, which naturally doesn't exist in the actual world. However, by using a dense bob and a light string, a real pendulum can be made to approach the properties of a simple pendulum.

tangent to the arc of swing at that point. We could draw such an inclined plane as part of a right triangle. The inclined plane would have a length L and would be at a height H above the horizontal line. The angle made by the inclined plane to the horizontal could be shown by ordinary geometry to be equal to the angle of displacement, and it, too, can be marked as θ.

As on page 19, the maximum gravitational force would have to be multiplied by the ratio of H to L, so the restoring force (F) would be equal to $mg(H/L)$. The ratio of H to L is usually thought of as the sine* of angle θ, and is symbolized as "sin θ." We can therefore represent the restoring force as:

$$F = mg(\sin \theta) \qquad \text{(Equation 8–6)}$$

The ratio of the restoring force to the displacement in the case of the swinging pendulum is therefore (combining Equations 8–5 and 8–6):

$$\frac{F}{D} = \frac{mg(\sin \theta)}{l\theta} \qquad \text{(Equation 8–7)}$$

Now the question is whether this ratio is a constant, as it must be if the swinging pendulum is to be considered an example of simple harmonic motion. The mass (m) of the bob and the

---

* The ratio of one side of a right triangle to another varies according to the size of the angles of the right triangle. For a given angle, these ratios are fixed, and each is given a name of its own. Since such ratios are studied in that branch of mathematics called "trigonometry" ("the measurement of triangles" is the meaning of the Greek term from which that expression is derived), such ratios are called *trigonometric functions*. Sines are an example of a trigonometric function. We don't have to go into these in detail. Suffice it to say that it is easy to obtain tables that will give the sine, or any of various other trigonometric functions, for angles of any size.

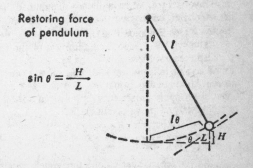

Restoring force
of pendulum

$$\sin \theta = \frac{H}{L}$$

length of the string (l) do not change as the pendulum swings, and the value of g is constant for any given point of the earth's surface, so the quantity mg/l may be considered constant. It remains only to determine whether the quantity $(\sin \theta)/\theta$ is likewise constant. If it is, we are set.

Unfortunately, the ratio is not constant. As we can easily determine, the sine of 30° is 1/2, while the sine of 90° is 1. The angle has tripled, in other words, while the sine of the angle has only doubled. This means that $(\sin \theta)/\theta$ is not a constant, that the restoring force of a pendulum is not proportional to the displacement, and that the swinging of a pendulum is *not* an example of simple harmonic motion.

Nevertheless, if the ratio $(\sin \theta)/\theta$ is not constant, it is nevertheless almost constant for small angles of 10° or less. Therefore, if the pendulum swings back and forth in moderate arcs, it is *almost* an example of simple harmonic motion.

In fact, for small angles $(\sin \theta)/\theta$ is not only constant, it is about equal to unity. For that reason (provided we remember that we are dealing with pendulums swinging through small arcs only), we can eliminate $(\sin \theta)/\theta$ in Equation 8–7 and write:

$$\frac{F}{D} \approx \frac{mg}{l} \qquad \text{(Equation 8–8)}$$

where the symbol $\approx$ signifies "is approximately equal to."

(You may wonder why we are willing to bother with an approximate equality when science should concern itself with exact relationships. The answer is that by being satisfied with an approximation, we can treat the pendulum as an example of simple harmonic motion and make certain other calculations quite simple, if not quite exact.)

For instance, we have already determined that the period (t) of an object undergoing simple harmonic motion is equal to $2\pi\sqrt{m/k}$ (see Equation 8–4).

The symbol k represents the ratio of the restoring force to the displacement for which, in the case of the pendulum, we have found a value in Equation 8–8, where it is set approximately equal to mg/l. Combining Equations 8–4 and 8–8 (and retaining the symbol for approximate equality) we can state that the period of a moderately swinging pendulum is:

$$t \approx 2\pi \sqrt{\frac{m}{mg/l}} \approx 2\pi \sqrt{\frac{l}{g}} \qquad \text{(Equation 8–9)}$$

As you see, the period of a moderately swinging pendulum is

independent of the mass of the bob and depends (at least to a very close approximation) upon the square root of the length of the string—as Galileo had determined by experiment back in the sixteenth century.

The presence of g, the acceleration due to gravity, is of great importance. If we solve Equation 8–9 for g, we obtain:

$$g \approx \frac{4\pi^2 l}{t^2}$$

(Equation 8–10)

This gives us a far easier method for measuring g than by trying to measure the velocity of free fall directly. The length of a pendulum is easily determined, and so is its period. The use of pendulums in Newton's time showed the manner in which g varied slightly with latitude and added experimental proof to Newton's suggestion that the earth was an oblate spheroid (see page 56).

Since the period of a moderately swinging pendulum is virtually constant, it can also be used to measure time. If a pendulum is connected to geared wheels in such a way that with each oscillation of the pendulum the wheel is pushed forward just one notch, the motion can then easily be scaled down in such a way as to push one pointer around a dial in exactly one hour (the minute hand) and another pointer around the same dial in exactly twelve hours (the hour hand). The addition of weights can keep the pendulum going so that the effects of friction and air resistance do not damp its motion to zero.

In his old age, Galileo had a vision of this application of his youthful discovery, but it was first brought to fruition by the Dutch scientist Christian Huygens (1629–1695) in 1673. Huygens even allowed for the imperfections of the pendulum. He showed how to take into account the fact that an actual pendulum is not a simple pendulum but has a bob of finite volume suspended from a string or rod of finite mass. He also showed that if a pendulum swung in a curve that was not the arc of a circle but the arc of a rather more complicated curve called a cycloid, the period would then be truly constant. He showed, furthermore, how the pendulum could be made to swing in such a cycloidal arc.

Since his time, ingenious methods have also been used to take into account the fact that the length of a pendulum (and therefore its period as well) changes slightly with changes in temperature.

Other examples of simple harmonic motion can also be used to measure time. Hooke (of Hooke's law) devised the "hairspring," a fine spiral spring which can be made to expand and contract in simple harmonic motion. The fine spring is driven by the uncoiling of a larger "main-spring" that is periodically tightened by mechanical winding. Such hair-springs are used in wristwatches, where there is obviously no room for a pendulum and where (even if room existed) the movements of the arm would throw a pendulum into confusion at once.

In recent years, the vibrations of the atoms moving within molecules in accordance with the rules of simple harmonic motion have been used to measure time. Such "atomic clocks" are far more regular and accurate than any clock based on supra-atomic phenomena can be.

CHAPTER 9

Liquids

So far I have assumed that the "bodies" which have been under discussion are *solid*—that is, that they are more or less rigid and have a definite shape. They resist any force tending to alter or deform that shape (though if the force increases without limit, a point is eventually reached where even the most rigid solid shape will deform or break). Solids behave all-in-a-piece, so if part of a solid moves, all of it moves, and in such a way as to maintain the shape.

There are bodies, however, which do not have a definite shape and do not resist deformation. If a stretch or shear, even a small one, is exerted upon them, they alter shape in response. In particular, they will respond to the force of gravity and alter their shape in such a way as to reduce their potential energy to a minimum. In response to gravity, such bodies will move downward and flatten out as much as possible; in so doing, they will take on the shape of any container in which they might be. If the container is open at the top and is tipped, or if an opening is made at the bottom, the material will pour out, under the influence of gravity, to take up a new position of still lower potential energy on the table-top, the floor or in a hole. It is this ability to pour or flow that gives such bodies the name of *fluids* (from a Latin word meaning "to flow").

Fluids fall into two classes. In one class, the downward force of gravity is paramount, so the fluid, while taking on the shape of the container, collects in the lowermost portions and does not necessarily fill it. Such fluids have a definite volume, if not a definite shape, and are called *liquids* (also from a Latin word meaning "to flow"). Water is, of course, the most familiar liquid.

In the other class of fluids, the downward force of gravity is countered by other effects to be discussed in later chapters. In this class there is a certain concentration toward the bottom of a container but not enough to notice under ordinary conditions. On the whole, such a fluid spreads itself more or less evenly through a confined space and has no definite volume of its own. Such fluids without either a definite shape or a definite volume are *gases.** Air is the most familiar gas.

I will take up each variety of fluid separately and will begin with liquids.

The weight of an object, as I explained earlier, is a downward force exerted by that object in response to the gravitational pull. In the case of solids, this force makes itself evident through whatever portion of its nether surface makes contact with another body. Since the nether surface is usually rough (even if only on a microscopic scale) the force is uneven, being exerted at those points where contact is actually made and not at others where contact is not made. For this reason, it is usually convenient to speak only of the total downward force exerted by a solid body, and this is indeed done when we speak of its weight.

In the case of a liquid, however, the contact between its nether surface and the object it rests upon is quite smooth and evenly distributed, so that all portions receive their equal share.† For fluids, therefore, it becomes convenient to speak of weight (or, more properly, force) per unit area. This force per unit area is termed *pressure*.

It is common to use as a unit of pressure "pounds per square

---

* The word "gas" was coined about 1600 by a Flemish chemist, Jan Baptista van Helmont (1577–1644), who supposedly derived it from the Greek word "chaos."

† If we get sufficiently submicroscopic, unevenness does show up, to be sure. This is because matter is not really continuous but is composed of discrete particles called atoms. We don't have to worry about this right now, but will consider it later (see page 143).

inch" (sometimes abbreviated "psi") where pounds are a unit of weight in this connection and *not* units of mass.

In the metric system the proper units of pressure are newtons per square meter in the mks system, and dynes per square centimeter in the cgs system. Since a newton equals 100,000 dynes and a square meter equals 10,000 square centimeters, 1 newton/m² is the equivalent of 100,000 dynes per 10,000 square centimeters, or 10 dynes/cm². Translating into metric units, one pound per square inch is equal to 6900 newtons/m², and one gram per square centimeter is equal to 98 newtons/m².

Suppose we consider a square centimeter of the bottom of a container filled with liquid to a height n. The pressure (dynes /cm²) depends on the weight of liquid resting on that square centimeter. The weight depends, in part at least, upon the volume of the column, one square centimeter in cross-sectional area and n centimeters high. The volume of that column is n cubic centimeters.

It does not follow, however, that in knowing the volume of a substance we also know its weight. It is common knowledge that the weight of a body of given volume varies according to the nature of the substance making up the body. We are all ready to admit, for instance, that iron is "heavier" than aluminum. By that, of course, is meant that a given volume of iron is heavier than the same volume of aluminum. (If we remove this restriction to equal volumes, we will be faced by the fact that a large ingot of aluminum is much heavier than an iron nail.)

For any object the weight per unit volume is its *density*, and in the metric system the units of density are usually expressed as grams (of weight) per cubic centimeter, or kilograms (of weight) per cubic meter. We should, therefore, say that iron is "denser," rather than "heavier," than aluminum.

If the height of a column of liquid resting upon a unit area determines its volume, and the density of that liquid gives the weight of a unit volume, then the total weight on the unit area, or pressure (p), is equal to the height of the liquid column (h) multiplied by its density (d):*

$$p = hd \qquad \text{(Equation 9-1)}$$

---

* This assumes that the density does not vary along the height of the column, and as far as liquids are concerned, any variation of density with depth is small enough to be ignored for small pressures. This will not be so for gases (see page 145).

The pressure of a liquid on the bottom of a container therefore depends only upon the height and density of the liquid, and not upon the shape of the container or the total quantity of liquid in the container. This means that the various containers shown in the accompanying figure, with bottoms of equal areas but with different shapes and containing different quantities of liquid, will have their bottoms placed under equal pressure.

It is easy to see that the container with the expanded upper portion ought to experience the same pressure at the bottom, for the weight of the additional liquid is clearly supported by the upper horizontal portion of the container. It may seem not at all logical, however, that the container with the contracted upper portion should also experience the same pressure at the bottom. The missing liquid (not present because of the contraction) has no weight to contribute to the pressure. How then does the pressure remain as great as if the missing liquid were there?

To explain that, we must realize that pressure is exerted differently in liquids as compared with solids. A solid resists the deforming influence of its own weight. A large pillar of marble may rest solidly on a stone floor and transmit a great deal of pressure to that floor, but it will itself remain unmoved under its own weight. The pillar will not, for instance, belly out in the middle,

Pressure and shape

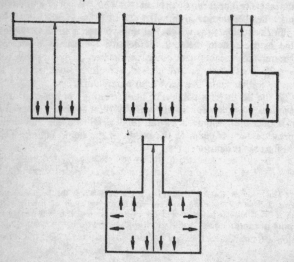

and if we place our hands on the side of the pillar we will be aware of no pressure thrusting out sideways.

Imagine, however, a similar pillar made of water. It could not remain in existence for more than a fraction of a second. Under the force of its own weight it would belly outward at every point and collapse. If a pillar of water is encased in a restraining cylinder of aluminum, the outward-bellying tendency of the water will evidence itself as a sidewise force. If a hole is punched in the aluminum cylinder, water will spurt out sidewise under the influence of that force. This same line of reasoning would show that a liquid would exert a pressure against a diagonally slanted wall with which it made contact.

A fluid, indeed, exerts pressure in all directions and particularly in a direction perpendicular to any wall with which it may make contact. The amount of pressure exerted at any given point depends upon the height of liquid above that particular point. Thus if a hole is punched in a cylindrical container of water, the liquid will spurt out with more force if the hole is near the bottom (with a great height of liquid above) than if it is near the top (with but a small height of liquid above).

In the container with a contracted upper portion, then, there is a pressure of the fluid up against the horizontal section, as indicated in the diagram. The amount of this pressure depends upon the height of the liquid above that horizontal section. By Newton's third law, the upper horizontal section exerts an equal pressure down upon the liquid. The downward pressure of the horizontal section is equal to that which would be produced by the missing liquid if it were there, and so the pressure at the bottom of the container remains the same.

Buoyancy

The generalization concerning pressure, made use of in the previous section, was first clearly stated by the French mathematician Blaise Pascal (1623–1662) and is therefore often referred to as *Pascal's principle*.

This can be expressed as follows: Pressure exerted anywhere on a confined liquid is transmitted unchanged to every portion of the interior and to all the walls of the containing vessel; and is always exerted at right angles to the walls.

This principle can be used to explain the observed fact that if a container of liquid contains two or more openings, to which are connected tubes of various shapes into which the liquid can rise, and if enough liquid is present in the container so that the level will rise into those tubes, the liquid will rise to the same height in each.*

To explain this, let us consider the case of a container with two openings and let us imagine a porous vertical partition dividing the container between the two openings. The pressure against the partition from the left would depend on the height of the liquid on the left, while the pressure from the right would depend on the height of the liquid on the right. If the liquid column is higher on the left, the pressure from the left is greater than that from the right, and there is a net pressure from left to right. Liquid is forced through the partition in that direction, so that the height of the liquid on the left decreases and that on the right increases. When both heights are equal there is no net pressure either left or right, and therefore no further motion.

This effect is part of folk knowledge, as is witnessed by the common saying that "water seeks its own level."

Notice that I am taking for granted here that liquids will move, or flow, in response to a force, and this is actually so. The laws of motion apply to fluids as well as to solids, and the study of mechanics includes, in its broad sense, forces and motions involving fluids as well as solids. However, it is quite common to restrict the use of the term "mechanics" to solid bodies. The mechanics of liquids is then given the special name *hydrodynamics* (from Greek words meaning "the motions of water"), and the mechanics of gases is called *pneumatics* (from the Greek word for "air"). These may be grouped together as *fluid mechanics*.

It is not only the weight of the liquid itself that can be trans-

* This is not strictly true because of capillary action. but this will be taken up later (see page 129).

Water finds its level

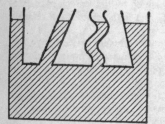
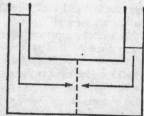

mitted to every part of the liquid as a pressure. Any applied force can be so transmitted.

For instance, suppose a liquid completely fills a container with two necks, each neck being stoppered by a movable piston which we can assume to be weightless. Suppose, furthermore, that the necks are of different width, so the piston in the larger neck has a cross-sectional area of 10 cm^2 while that in the smaller one has a cross-sectional area of only 1 cm^2.

Now imagine that a force of one dyne is exerted downward on the smaller piston. Since the area of the smaller piston is 1 cm^2, the pressure upon it as a result of the applied force is 1 dyne/cm^2. In accordance with Pascal's principle, this pressure is transmitted unchanged through the entire body of liquid and, perpendicularly, to all the walls. It is transmitted, in particular, perpendicularly to that portion of the wall represented by the larger piston. As the small piston moves downward, then, the large piston moves upward.

The upward pressure against the larger piston must be the same as the downward pressure against the smaller piston, 1 dyne /cm^2. The area of the larger piston is, however, 10 cm^2. The total force against the larger piston is therefore 1 dyne/cm^2 multiplied by 10 cm^2, or 10 dynes. The total force has been multiplied tenfold and the weight which the original force would have been capable of lifting has also been multiplied tenfold. It is by "hydraulic presses" based on this effect that heavy weights can be lifted with an expenditure of but a reasonable amount of force.

Are we in this way getting something for nothing? Not at all!

Suppose we press down on the small piston (1 cm^2 in area)

Hydraulic press

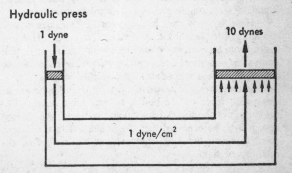

1 dyne

10 dynes

1 dyne/cm^2

and make it move a distance 1 cm. The volume of liquid it has displaced is 1 cm² multiplied by 1 cm, and this comes to one cubic centimeter (1 cm³). The larger piston (10 cm² in area) can only move upward a sufficient distance to make room for the displaced 1 cm³ of liquid. The distance required is 1 cm³ divided by 10 cm², or 0.1 cm. Thus the situation is the same as it was for the lever (see page 89). The force has been multiplied tenfold, yes, but the distance through which the force has been exerted has been reduced to one-tenth. The total work (force times distance) obtained from the hydraulic press is the same, if we neglect such things as friction, as the total work put into it.

The pressure of a liquid will be transmitted not only to the walls of a container but also (perpendicularly) to the surfaces of any solid object within the liquid. Imagine a cube of iron suspended in liquid so that the top and bottom surface of the cube are perfectly horizontal and the other four surfaces are perfectly vertical. The pressure against each of the four vertical surfaces depends on the height of liquid above them, which is the same for all. For the vertical surfaces, then, we have equal pressures arranged in opposing pairs. There is, consequently, no net sideways pressure in any direction.

But what if we consider the two horizontal surfaces, the one on top and the one on bottom? It is clear that there is a greater height of liquid above the lower surface than above the upper one. There is therefore a comparatively great upward pressure against the lower surface and a comparatively small downward pressure against the upper surface. As a result, a net upward force is exerted by the liquid upon the submerged object. (This is most easily reasoned out in the case of the solid cube, but it can be shown to hold for a solid of any shape or, for that matter, for a submerged drop of liquid or bubble of gas.) This upward force of liquids against submerged objects is called *buoyancy*.

How large is this buoyant force? Consider a solid body dropping into the liquid contents of a vessel. The solid must make room for its own volume by pushing aside, or displacing, an equivalent volume of liquid, and the liquid level in the vessel rises sufficiently to accommodate that displaced volume.

It therefore follows that the submerging solid is exerting a downward force on the liquid, a force large enough to balance the weight of the solid's own volume of liquid. By Newton's third law, it is to be expected that the liquid will in turn exert an upward force on the solid equivalent to the weight of that same quantity of liquid.

The original weight of the submerged body is equal to its volume (V) times its density (D). The weight of the displaced liquid is equal to its volume (which is the same as the volume of the submerged solid, and hence also V) times its density (d). The weight of the body after submersion (W) is equal to its original weight minus the weight of the displaced water:

$$W = VD - Vd$$ (Equation 9–2)

Solving for D, the density of the submerged solid, we have:

$$D = \frac{W + Vd}{V}$$ (Equation 9–3)

The weight of the immersed body (W) can be directly measured, the volume of the displaced fluid (V) is obtained at once from the rise in water level and the cross-sectional area of the container, and the density of the fluid (d) is also easily measured. With this data in hand, the density of the immersed body can be calculated easily from Equation 9–3.

This method of measuring density was first made use of by the Greek mathematician Archimedes in the third century B.C. The story is that King Hiero of Syracuse, having received a gold crown from the goldsmith, suspected graft. The goldsmith had, the king felt, alloyed the gold with cheaper silver and had pocketed the difference. Archimedes was asked to tell whether this had been done, without, of course, damaging the crown.

Archimedes knew that a gold-silver alloy would have a smaller density than would gold alone, but he was at a loss for a method of determining the density of the crown. He needed both its

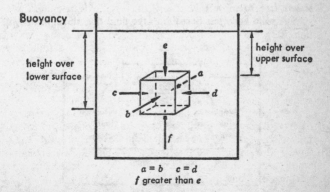

Buoyancy

height over lower surface

height over upper surface

$a = b$ $c = d$
f greater than e

weight and its volume for that, and while he could weigh it easily enough, he could not estimate the volume without pounding it into a cube or sphere or some other shape for which the volume could then be worked out by the geometry of the time. And pounding the crown would have been frowned on by Hiero.

The principle of buoyancy is supposed to have occurred to Archimedes when he lowered himself into a full bathtub and noted the displaced water running over the sides. He ran naked through the streets of Syracuse (so the story goes) shouting, "Eureka! Eureka!" ("I've got it! I've got it!"). By immersing the crown in water and measuring the new weight together with the rise in water level, and then doing the same for an equal weight of pure gold, he could tell at once that the density of the crown was considerably less than that of the gold; the goldsmith was suitably punished. The principle of buoyancy is sometimes called *Archimedes' principle* as a result.

If an immersed body has a greater density than that of the fluid in which it is immersed, then D is greater than d, and VD is naturally greater than Vd. From Equation 9–2, we see that in that case the weight (W) of the immersed body must be a positive number. The weight of the body is decreased, but it is still larger than zero and falls through the fluid. (Thus a solid iron or aluminum object will fall through water.)

However, if the immersed body has a smaller density than that of the fluid, then D is smaller than d, VD is smaller than Vd, and the immersed body has a weight that is a negative number. With a negative weight (so to speak), it moves upward rather than downward in response to a gravitational field. (Thus a piece of wood or a bubble of air submerged in water will "fall upward" if left free to move.)

A solid body less dense than the fluid that surrounds it will

Floating

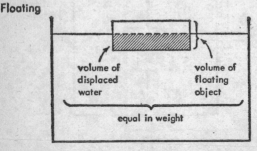

volume of displaced water   volume of floating object

equal in weight

float, partly submerged, on the surface of the fluid under conditions where the weight of water it displaces is equal to its own original weight; in such a case, its weight in water is zero, and it neither rises nor falls. The solid body floats when it has displaced just enough water (less than its own volume) to equal its own original weight.

It is not to be supposed though that because a steel ship floats the density of steel is less than that of water. It is not the steel of the ship alone that displaces water. The ship is hollow, and as it sinks into the water the enclosed air displaces water just as the steel does. The density of the steel-plus-enclosed-air is less than the density of water, though the density of steel alone most certainly is not, and so the steel ship floats.

The force of buoyancy, by the way, is not a matter of calculations and theory alone; it can easily be felt. Lift a sizable rock out of the water and its sudden gain in weight as it emerges into the air can be staggering. Float a sizable block of wood on a water surface and try to push it down so that it will be completely submerged, and you will feel the counterforce of buoyancy most definitely.

Cohesion and Adhesion

Solids, as I said at the beginning of the chapter, act all-in-a-piece. Each fragment of a solid object clings firmly to every other fragment, so if you seize one corner of a rock and lift, the entire substance of the rock rises. This sticking-together is called *cohesion* (from Latin words meaning "to stick to").

Fluids have nothing like the kind of cohesion that exists in solids. If you dip your hand into water and try to lift a piece of it in the hopes that the entire quantity will rise out of its con-

Surface tension

tainer, you will only get your fingers wet. Nevertheless, one should not conclude from this that the force of cohesion in liquids is completely absent. The force is much smaller in most liquids than in solids, but it is not entirely zero. This is most clearly seen at the surface of a liquid.

In the body of a liquid, even a short way below the actual surface, a given portion of the liquid is attracted by cohesive forces in all directions equally by the other portions of the liquid that surround it. There is no net unbalanced force in any particular direction.* At the surface of the liquid, however, the cohesive forces are directed only inward toward the body of the liquid and not outward where there is no liquid to supply cohesive forces. (Most often, there is only air on the other side of the liquid surface, and the attractive forces between air and the liquid are so small that they can be ignored.) The resultant of this semisphere of cohesive forces about a particle of liquid in the surface is a net inward force exerted perpendicularly from that surface.

To keep liquid in the surface against this inward force requires work, so the surface represents a form of energy of position or potential energy. This particular form is usually called *surface energy*.

Such surface energy is distributed over an area of surface, so its units are those of work per area. In the mks system this would be joules per square meter (joules/m^2), and in the cgs system it would be ergs per square centimeter (ergs/cm^2). In the case of surface energy, the cgs system is the more convenient and is generally used. One erg is equal to one dyne-cm or 1 g-cm^2/sec^2 so 1 erg/cm^2 is equal to 1 (g-cm^2/sec^2)/cm^2. If we cancel one of the centimeter units, this becomes 1 (g-cm/sec^2)/cm, or 1 dyne/cm, and as a matter of fact the units of surface energy are most often presented with the last-named unit, dynes per centimeter.

Left to itself, surface energy reduces to a minimum in a way analogous to that in which gravitational potential energy is reduced whenever a ball high in the air falls to the ground, or in which a column of liquid flattens and spreads out if the container is broken. A small quantity of liquid suspended in air will take up the shape of a sphere, for a sphere has for its volume the smallest area of surface, therefore, surface energy is then reduced to a

---

* In solids, the various particles constituting the substance are lined up in fixed and orderly positions (rather than moving about freely as in liquids). For that reason, the cohesive forces between neighboring particles in solids are oriented in definite directions and are very noticeable.

minimum. Such a sphere of liquid is, however, distorted into a "tear-shaped" object by the unbalanced downward pull of gravity. If it is falling through air, as a raindrop does, for instance, it will be flattened at the bottom through the upward force of air resistance. The smaller the droplet of liquid, the smaller the relative effects of gravity and air resistance, and the more nearly spherical it is. Soap bubbles are hollow, liquid structures that are so light for their volume (because of enclosed air) that the forces of gravity (unusually low in this case) and of air resistance (unusually high) cancel each other. Soap bubbles therefore drift about slowly and show themselves to be virtually perfect spheres.

A sizable quantity of liquid flattens out. The necessity of minimizing the gravitational potential energy rises superior to that of minimizing the surface potential energy, and the surface of an undisturbed pail of water (or pond of water) seems to be a plane. Actually, it is a segment of a sphere, but a large one; one that has a radius equal to that of the earth. Look at the Pacific Ocean on a globe of the earth and you will see that its surface almost forms a semisphere.

If energy in any form is added to a liquid, some may well go into increasing the surface energy by extending the surface area beyond its minimum. Thus, wind will cause the surface of an ocean or lake to become irregular and therefore increase in surface area. The surface in a glass of water will froth if the glass is shaken.

Because the surface is stretched into a larger area by such an input of energy, and because it pulls back to the minimum when the energy input ceases, the analogy between the liquid surface and an elastic skin under tension (a very thin film of stretched rubber, for instance) is unmistakable. The surface effects are therefore frequently spoken of as being caused by *surface tension*, rather than by surface energy.

The same sort of cohesive forces that act to hold different portions of a liquid together, via surface tension, act also to hold a portion of a liquid in contact with a portion of a neighboring solid. In the latter case, where the attractive force is between solid and liquid (unlike particles) rather than between liquid and itself (like particles), the phenomenon is called *adhesion* (also, like cohesion, from Latin words meaning "to stick to"). Adhesive forces may be as great as, or even greater than, cohesive forces. In particular, the adhesion of water to clean glass is greater than the cohesion of water to itself.

This has an effect on the shape of the liquid surface of water in a glass container. Where the water meets the glass, the attraction of the glass for water is large enough to overcome water's cohesive forces. As a result, the water surface rises upward so as to increase the water-glass contact (or "interface") as much as possible at the expense of the weaker water-water forces. If there were no countering forces, water would rise to the top of the container and over. However, there is the countering force of gravity. There comes a point where the weight of the raised water, added to the cohesive forces of water, just balances the upward pull of the adhesive forces, and a point of equilibrium is reached after the water level has been raised by a moderate degree.

If the container is reasonably wide, this upward-bending of the surface is restricted only to the neighborhood of the water-glass contact. The water surface in the interior remains flat. Where the container is a relatively narrow one, however, the surface of the liquid is all in the region of water-glass contact, and the liquid surface is then nowhere plane; instead it forms a semisphere bending down to a low point in the center of the tube. Viewed from the side, the surface resembles a crescent moon and, indeed, it is spoken of as a *meniscus* (Greek for "little moon").

Cohesive forces may well be larger than adhesive forces in particular cases. For instance, the cohesive forces in liquid mercury are much larger than those in water; they are also larger than the adhesive forces between mercury and glass. If we look at mercury in a glass tube, we see that at the interface where mercury meets glass, the mercury pulls away from the glass, reducing the mercury-glass interface. The mercury meniscus in such a tube bends downward at the edges and rises to a maximum height at the center of the tube. The same is true even for water if the glass container has a coating of wax, since the adhesive forces between water and wax are less than the cohesive forces within water.

If water is spilled onto a flat surface of glass, it will spread out into a thin film so as to make the greatest possible contact, adding to the total adhesive force at the expense of the weaker cohesive force. The water, in other words, wets the glass. Mercury, however, when spilled on glass (or water on a waxed surface), makes as little contact with the glass as possible, drawing itself into a series of small gravity-distorted spheres, and adding to the total cohesive force at the expense of the weaker adhesive force. Mercury does not wet glass, and water does not wet wax. In all these events, the effect is to reduce the total surface energy (that of the liquid/air interface plus that of the liquid/solid interface) to a minimum.

Where a water-containing tube attached to a water reservoir is narrow, the rise in water level brought about by the upward force of adhesion is considerable, and the water rises markedly above its "natural level" (see page 120).

It is possible to calculate what the raised height (h) of the water level must be in a particular tube. Adhesion is a form of surface tension (which we can represent as the Greek letter "sigma," σ) acting around the rim of the circle where water meets the glass of the tube. This circle has a length of $2\pi r$, where r is the radius of the tube. The total upward force brought about by adhesion is therefore the surface tension of the water-glass interface, σ dynes/cm, multiplied by the length of the circle where water and glass meet, $2\pi r$ cm, so that the total force is $2\pi r\sigma$ dynes.

Countering this upward force is the downward force of gravitation, which is equal to the weight (mg dynes, see page 54) of the raised water. The mass of the column of water raised by adhesion is equal to its volume (v) times its density (d). Substituting vd for m, we see that the weight of the water is vdg dynes. Since the raised column of water in the tube is in the form of a cylinder, we can make use of the geometrical formula for the volume of a cylinder and say that the volume of the raised water is equal to the height of the column (h) multiplied by the cross-sectional area (πr^2), where r is the radius of the column. Substituting $\pi r^2 h$ for v, we see that the weight of the water is $\pi r^2 hdg$ dynes.

When the water in the narrow tube has been raised as high as it will go, the upward adhesive force is balanced by the downward gravitational force, so we have:

$$2\pi r\sigma = \pi r^2 hdg \qquad \text{(Equation 9–4)}$$

Solving for h:

$$h = \frac{2\sigma}{rdg} \qquad \text{(Equation 9–5)}$$

The acceleration due to gravity (g) is fixed for any given point on the earth; and for any particular liquid, the surface tension (σ) and the density (d) are fixed for the particular conditions of the experiment. The important variable is the radius of the tube (r). As you see, the height to which a column of water is drawn upward in a narrow tube is inversely proportional to the radius of the tube. The narrower the tube, the greater the height to which the liquid is lifted. Consequently, the effect is most noticeable in tubes (natural or artificial) of microscopic width. These are capillary tubes (from a Latin expression meaning "hair-like"), and the rise of columns of water in such tubes is called *capillary action*. It is through capillary action that water rises through the narrow interstices of a lump of sugar or a piece of blotting paper, and it is at least partly through capillary action that water rises upward through the narrow tubes within the stems of plants.

Again, if we know the value of the density of a liquid and the extent of its rise in a tube of known radius (both rise and radius being easily measured), it follows that since the value of *g* is also known, the value of the surface tension (σ) can be calculated from Equation 9–5.

In the case of mercury, where the adhesive forces with glass are exerted downwards, the level is pulled below the "natural level." The degree to which the level is lowered is increased as the radius of the tube is decreased.

Viscosity

We are accustomed to the notion of friction as a force that is exerted opposite to that which brings about motion when one solid moves in contact with another. Such friction tends to slow,

Capillary action

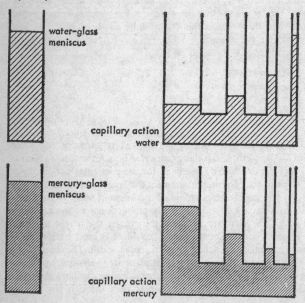

water-glass meniscus

capillary action water

mercury-glass meniscus

capillary action mercury

and eventually stop, motion unless the propulsive force is vigorously maintained.

There is also friction where a solid moves through a fluid, as when a ship plows through water. For all that water seems so smooth and lacking in projections to catch at the ship, the ship once set in motion will speedily come to a halt, its energy absorbed in overcoming the friction with the water, unless the propulsive force is vigorously maintained there, too.

This friction arises from the fact that it is necessary to expend energy to pull the water apart against its own cohesive forces in order to make room for the ship or other object to pass through. The energy expended varies with the shape of the object moving through the fluid. If the fluid is pulled apart in such a way as to force it into eddies and other unevennesses of motion (*turbulence*), the energy expended is multiplied and the motion stops the sooner; to prevent a stop the propulsive force must be increased. If, instead, the fluid is pulled apart gradually by the forward edge of the moving object and allowed to come together even more gradually behind, so turbulence is held at a minimum, the energy expended is reduced considerably, and the force required to maintain motion is likewise reduced. A "streamlined shape" consisting of a bluntly curving fore and a narrowly tapering rear is the "teardrop" shape of water-drops falling through air, and of fish, penguins, seals, and whales moving through water. It is used in human devices, too, where motion through a liquid medium with maximum economy is desired—such use was enforced by hit-and-miss practice long before it was explained by theory.

The friction between a moving solid and a surrounding liquid increases with velocity. Thus, an object falling through water is accelerated by the gravitational pull against the resistance of friction with the water. However, as the velocity of the falling body increases, the resisting friction increases, too; the force of gravity, of course, remains constant. Eventually, the resisting force of friction increases to the point where it balances the force of gravity, so acceleration is then reduced to zero. Once that happens, the body falls through the liquid at a constant *terminal velocity*.

We ourselves are easily made aware of the friction of solids moving through liquids. Anyone trying to walk while waist-deep in water cannot help but be conscious of the unusual consumption of energy required and of the "slow-motion" effect.

The friction makes itself evident even when the liquid itself is the only substance involved. When a liquid moves, it does not move all-in-one-piece as a solid does. Instead, a given portion will move relative to a neighboring portion, and an "internal friction" between these two portions will counter the motion. Where the cohesive forces that impose this internal friction are low, as in water, we are not ordinarily very conscious of this. Where they are high, as in glycerol or in concentrated sugar solutions, the fluid pours slowly; so slowly indeed that, accustomed as we are to the comparatively rapid water flow, we tend to grow impatient with it. The internal friction is higher at low temperatures than at high temperatures. The folk-saying "as slow as molasses in January" points up our impatience.

A slowly-pouring liquid is said to be "viscous," from the Latin word for a sticky species of birdlime that had this property. The internal friction that determines the manner in which a liquid will pour is called the *viscosity*. There are liquids that are so viscous that the pull of gravity is not sufficient to make them flow against the strong internal friction. Glass is such a liquid and its viscosity is such that it seems a solid to the ordinary way of thinking.*

To consider the measurement of viscosity, imagine two parallel layers of liquid, each in the form of a square of a given area *a* and separated by a distance *d*. To make one of these squares move with respect to the other at velocity *v* against the resisting internal friction requires a force *f*. It turns out that the relationship among these properties can be expressed by the following equation:

$$\frac{fd}{va} = \eta \qquad\qquad \text{(Equation 9–6)}$$

where η (the Greek letter "eta") is a constant at a given temperature and represents the measure of viscosity.

The unit of viscosity can be determined from Equation 9–6. The expression *fd* in the numerator of the fraction in Equation 9–6 represents force multiplied by distance, or work. The unit of work in the cgs system is dyne-cm or $gm-cm^2/sec^2$. The expression *va* in the denominator of the fraction represents volume (centimeters per second) multiplied by area (square centimeters). The unit of *va*, therefore, is $(cm/sec)(cm^2)$ or cm^3/sec.

---

* That glass is not a solid, despite its seeming so, is evidenced by its lack of certain properties characteristic of solids. Glass does not have a crystalline structure, for instance, or a sharp melting point. Even so, the case of glass is evidence enough that the distinction between a solid and a liquid is not as clear-cut as might be expected from the most common examples of either. Indeed, most differences and distinctions in science are artificial human conventions imposed on a very complicated universe, and such distinctions cannot help but become fuzzy if viewed with sufficient attention to detail.

To get the unit of viscosity in the cgs system, we must therefore divide the unit of *fd* by that of *va*. It turns out that (gm-cm²/sec²)/(cm³/sec) works out by ordinary algebraic manipulation to gm/cm-sec, or grams per centimeter-second. One gm/cm-sec is defined as one *poise* in honor of the French physician Jean Louis Marie Poiseuille (1799–1869), who in 1843 was the first to study viscosity in a quantitative manner. (As a physician, he was primarily interested in the manner in which that viscous fluid, blood, moved through the narrow blood-vessels.)

The poise is too large for convenience in dealing with most liquids, so the *centipoise* (one-hundredth of a poise) and even the *millipoise* (one-thousandth of a poise) are commonly used. Thus, the viscosity of water at room temperature is just about one centipoise. At the same temperature, the viscosity of diethyl ether (the common anesthetic) is 0.23 centipoises, or 2.3 millipoises, while the viscosity of glycerol is about 1500 centipoises, or 15 poises.

The motion of a fluid has an effect upon its pressure. Imagine a column of water flowing through a horizontal tube of fixed diameter. The water is under pressure or it would not be moving, and the pressure (force per unit area) is the same at all points, for the water is flowing at the same velocity at all points. This could be demonstrated if the pipe were pierced at intervals and a

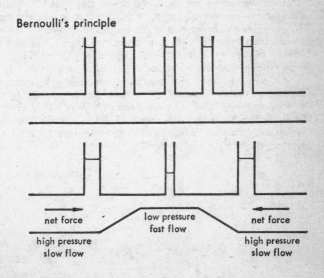

Bernoulli's principle

net force — high pressure slow flow — low pressure fast flow — net force — high pressure slow flow

tube inserted into each orifice. The water would rise to the same level in each tube.

But suppose the pipe had a constricted area in the middle. The same volume of water would have to pass through the constricted area in a given time as would have to pass through an equal length of unconstricted area. If that were not the case, water would pile up at the entrance to the constriction, which, of course, it does not. (If the constriction were narrow enough to prevent flow altogether, flow would stop, and the volume of water passing through a given section would be 0 cm³/sec in the constricted and unconstricted areas alike.)

But in order for the same volume of water to pass through the constricted and unconstricted areas in a given time the flow of water must be more rapid through the constricted area (just as the wide slowly-flowing river becomes a tumbling torrent when passing through a narrow gorge). Since the velocity of water increases as it enters the constricted area, it is subject to an acceleration, and this must be brought about by a force. We can most easily find such a force by supposing a difference in pressure. If the pressure in the unconstricted portion is greater than that in the constricted portion, then there is a net force from the unconstricted portion (high pressure) toward the constricted portion (low pressure), and the liquid is indeed accelerated as it enters the constriction.

Furthermore, when the liquid leaves the constriction and enters a new unconstricted area, its velocity must decrease again. This involves an acceleration again, and there must be a force in the direction opposite to the flow in order to bring about such a slowing of velocity. However, if the new unconstricted area is a region of high pressure again, such a force can be accounted for.

In short, it can be concluded as an important generalization that the pressure of a liquid (or a fluid, generally) falls as its velocity increases. This is called *Bernoulli's principle*, after the Swiss mathematician Daniel Bernoulli (1700–1782), who was the first to study the phenomenon in 1738 and who, on that occasion, invented the term "hydrodynamics."

CHAPTER **10**

Gases

Density

The properties of liquids, described in the previous chapter, are important in connection with a fundamental question concerning the ultimate composition of matter, a question that was of great interest to scientists as long ago as the time of the ancient Greeks.

Matter can be subdivided indefinitely, as far as the eye can see. A piece of paper can be torn in half, in quarters, and in eighths—and still remain paper. A drop of water can be divided into two smaller drops or into four still smaller drops—and still remain water. Can such a subdividing process be continued forever? Is matter continuous even to ultimate smallness? There was no way in which the ancient thinkers could test this in actual practice, and they resorted to logical arguments based on what they considered first principles.

Some, notably Democritus of Abdera (fifth century B.C.), maintained that matter could not be subdivided forever, but that eventually a small portion was reached that could not possibly be broken down further. This he called "atomos" (meaning "uncuttable"), and we now speak of his views as representing *atomism*, or an *atomic theory*.

Other Greek philosophers, notably Aristotle, argued

against this notion, however, adducing reasons that made the idea of atoms seem illogical. By and large, the non-atomistic view won out and remained the prevalent belief of scientists for two thousand years.

If one confined oneself to the study of the properties of solids, one could scarcely help but be sympathetic to the Aristotelian view, for there is nothing about a solid that would make it seem logical to consider it to be composed of a conglomeration of small particles. If it were, we would have to suppose the particles to be stuck firmly together, since solids acted all-in-one-piece. And if we are going to suppose that particles are stuck firmly together, why not discard the particles altogether and suppose the solid to be all one piece of continuous matter in the first place?

Where liquids are concerned, the situation is quite different. By the very fact that liquids do not move all-in-one-piece, it might reasonably be suggested that they are composed of separate particles. A mass of tiny metal spheres or a heap of powder would take on the overall shape of any container in which they were placed, and they would pour as a fluid would. If the particles were rather sticky, they would pour like a viscous fluid.

In fact, many of the properties of liquids could be explained nicely by supposing them to consist of sub-submicroscopic particles which attract each other somewhat. Surface tension could be explained in this fashion, for instance.

However, all the properties that make liquids suggest atomism more effectively than solids do, are further intensified in gases. And in actual fact, it was the study of gases through the seventeenth and eighteenth centuries that finally forced scientists to reverse the early decision in Aristotle's favor and to take up again, at the start of the nineteenth century, the long-discarded view of Democritus.

Gases differ from liquids most clearly and obviously, perhaps, with respect to density. In comparison with liquids, gases are thin and rarefied.

The density of water is 62.43 pounds per cubic foot in the English system. Using metric units, it is one gram per cubic centimeter (1 gm/cm^3)* in the cgs system and a thousand kilograms per cubic meter $(5,000 \text{ kg/m}^3)$ in the mks system. The least dense liquid at room temperature or below is liquid hydrogen, while the densest is mercury. The former has a density of 0.07 gm/gm^3, the

---

* This is no coincidence. In setting up the metric system in the 1790's, the French originators defined the gram as the weight of a cubic centimeter of water under set conditions of temperature.

latter a density of 13.546 gm/cm³. (At elevated temperatures some metals such as platinum would melt to liquids with densities as high as 20 gm/cm³.)

The density of solids falls for the most part within this range, too. The lightest solid, solid hydrogen, has a density of 0.08 gm/cm³, while the heaviest, the metal osmium, has one of 22.48 gm/cm³.

Such densities can be expressed in allied fashion as *specific gravity*, a term dating back to the Middle Ages. Specific gravity may be defined as the ratio of the density of a substance to the density of water. In other words, if the density of mercury is 13.546 gm/cm³ and that of water is 1 gm/cm³, then the specific gravity of mercury is (13.546 gm/cm³)/(1 gm/cm³), or 13.546.

Because the density of water is 1 gm/cm³, the specific gravity comes out numerically equal to the density in the cgs system, but you must not be misled by this apparent equality, for there is an important difference in the matter of units. In dividing a density by a density, the units (gm/cm³, in the case cited in the previous paragraph) cancel, so that the figure for specific gravity is a dimensionless number.

The units cancel when specific gravity is calculated, no matter what system of units is used for the densities. In the mks system, the densities of mercury and water are, respectively, 13,546 kg/m³ and 1,000 kg/m³. By taking the ratio, the specific gravity of mercury is 13.546, as before. In the English system of units, the densities of mercury and water are 845.67 pounds per cubic foot and 62.43 pounds per cubic foot, and the ratio is still 13.546.

The convenience of a dimensionless number is just this then: it is valid for any system of units.

The specific gravity of gases is much less than that of either liquids or solids. The most common gas, air, has a specific gravity of 0.0013 under ordinary conditions. The lightest gas, hydrogen, has under ordinary conditions a specific gravity of 0.00009. An example of a very dense gas is the substance uranium hexafluoride, which is a liquid at ordinary temperatures, but if heated gently is converted into a gas with a specific gravity of 0.031.

Thus, under ordinary conditions, even the densest gases are less than half as dense as even the least dense liquids or solids, while a common gas such as air has only about 1/700 the density of a common liquid such as water, and only about 1/2000 the density of common solids such as the typical rocks that make up the earth's crust.

Gas Pressure

Gases share the fluid properties of liquids but in an attenuated form, as is to be expected considering the difference in density. For instance, gases exhibit pressure as well as liquids do, but gas pressure is considerably smaller for a given height of fluid. A column of air a meter high will produce a pressure at the bottom only 1/700 that of a column of water a meter high.

Nevertheless we live at the bottom of an ocean of air many miles high. Its pressure should be considerable and it is; it is equivalent to the pressure produced by a column of water ten meters high. This pressure was first measured by the Italian physicist Evangelista Torricelli (1608–1647) in 1644.

Torricelli took a long tube, closed at one end, and filled it with mercury. He then upended it in a dish of mercury. The mercury in the tube poured out of the tube, of course, in response to the downward pull of gravitational force. There was a counterforce, however, in the form of the pressure of the atmosphere against the mercury surface in the dish. This pressure was transmitted in all directions within the body of the mercury (Pascal's principle, see page 119), including a pressure upward into the tube of mercury.

As the mercury poured out of the tube, the mass of the column, and therefore the gravitational pull upon it, decreased until it merely equalled the force of the upward pressure due to the atmosphere. At that point of balancing forces, the mercury no longer moved. The mercury column that remained exerted a pressure (due to its weight) that was equal to the pressure of the atmosphere (due to its weight). The total weight of the atmosphere is, of course, many millions of times as large as the total weight of the mercury, but we are here concerned with pressure which, be it remembered, is weight (or force) per unit area.

It turns out that the pressure of the atmosphere at sea level is equal to that of a column of mercury just about 30 inches (or 76 centimeters) high; Torricelli had, in effect, invented the barometer. Air pressure is frequently measured, particularly by meteorologists, as so many *inches of mercury* or *centimeters of mercury*, usually abbreviated in Hg or cm Hg, respectively.* It is natural to set 30 in. Hg or 76 cm Hg equal to 1 *atmosphere*. A millimeter of mercury (mm Hg) has been defined as 1 *torricelli*, in honor of the physicist, so one atmosphere is equal to 760 torricellis.

---

* "Hg" is the chemical symbol for mercury.

Air pressure may also be measured as weight per area. In that case, normal air pressure at sea level is 14.7 pounds of weight per square inch, or 1033 grams of weight per square centimeter (gm[w]/cm^2). Expressed in the more formal units of force per area, one atmosphere is equal to 1,013,300 dynes/cm^2. One million dynes per square centimeter has been set equal to one *bar* (from a Greek word for "heavy"), so one atmosphere is equal to 1.0133 bars.

Naturally, if it is the pressure of the atmosphere that balances the pressure of the mercury column, then when anyone carrying a barometer ascends a mountain, the height of the column of mercury should decrease. As one ascends, at least part of the atmosphere is below, and what remains above is less and less. The weight of what remains above, and therefore its pressure, is lower and so is the pressure of the mercury it will balance.

This was checked in actual practice by Pascal in 1658. He sent his brother-in-law up a neighboring elevation, barometer in hand. At a height of a kilometer, the height of the mercury column had dropped by ten percent, from 76 centimeters to 68 centimeters.

Furthermore, the atmosphere is not evenly distributed about the earth. There is an unevenness in temperature that sets up air movements that result in the piling up of atmosphere in one place at the expense of another. The barometer reading at sea level can easily be as high as 31 in. Hg or as low as 29 in. Hg. (In the center of hurricanes, it may be as low as 27 in. Hg.) These "highs" and "lows" generally travel from west to east, and their movements can be used to foretell weather. The coming of a high (a rising barometer) usually bespeaks fair weather, while the coming of a low (a falling barometer) promises storms.

For all that air pressure is sizable in quantity (the value, 14.7 pounds of weight per square inch, is most easily visualized by a person used to the common measurements in the United States) it goes unnoticed by us. For thousands of years, men considered air to be weightless. (We still say "as light as air" or "an airy nothing.")

The reason for this is that air exerts its pressure in all directions, as all fluids do. An empty balloon, although supporting the full pressure of miles of air, will rest with its mouth open and its walls not touching, for the air within it has an outward pressure equal to the inward pressure of the air outside. Place the balloon in your mouth, however, and suck out the air within so that the inward push of the air outside is no longer balanced. Now the walls of the balloon will be pushed hard together.

The same factors apply to human beings. The air in our lungs, the blood in our veins, the fluid in our bodies (and living tissue is essentially a thick, viscous fluid) is generally at air pressure and delivers a pressure outward equal to that of the atmosphere inward. The net pressure exerted on us is zero, and we are therefore unaware of the weight of the air.

If we submerge ourselves in water, the pressure from without rapidly increases, and it cannot be matched by pressure from within without damage to our tissues. It is for this reason that an unprotected man, such as a skin diver, is severely limited in the depth to which he can penetrate, regardless of how well equipped with oxygen he may be. On the other hand, forms of life adapted to the deeps exist at the extremest abyss of the ocean, where the water pressure is over a thousand atmospheres. Those life forms are as unaware of the pressure (balanced as it is from within), and as unhampered by it, as we are by air pressure.

Once it was recognized that air had weight and produced a pressure, it was also quickly recognized that this could easily be demonstrated provided it were *not* balanced by an equal pressure from within. In other words, it seemed desirable to be able to remove the air from within a container, producing a *vacuum* (a Latin word meaning "empty") so that the air pressure from without would remain unbalanced by any appreciable pressure from within. Torricelli had formed the first man-made vacuum (inadvertently) when he had upended his tube of mercury. The column of mercury, as it poured out, left behind a volume of nothingness (except for thin wisps of mercury vapor), and this is still called a "Torricellian vacuum."

Just a few years later, in 1650, the German physicist Otto von Guericke (1602–1686) invented a mechanical device that little by little sucked air out of a container. This enabled him to form a vacuum at will and to demonstrate the effects of an unbalanced air pressure. Such air pressure would hold two metal hemispheres together against the determined efforts of two eight-horse teams of horses (whipped into straining in opposite directions) to pull them apart. When the air was allowed to enter the hemispheres once more, they fell apart of their own weight.

Again, air pressure gradually forced a piston into a cylinder being evacuated, even though fifty men pulled at a rope in a vain attempt to keep the piston from entering.

In other respects, too, a gas like air has fluid properties in an attenuated form. It exhibits buoyancy, for instance. We ourselves displace a volume of air equal to our own volume, and the effect is to cause a 150-pound man to weigh some three ounces less than he would in a vacuum. This is not enough to notice ordinarily, of course, but for objects of very low densities the effect is very noticeable.

This is particularly true for substances (such as certain gases) that are lighter than air. Hydrogen gas, for instance, has only 1/14 the density of air. In consequence, hydrogen penned within a container is subjected to an upward force like that of wood submerged in water (see page 124). If the container is light enough, it will be carried upward by this upward force. If enough hydrogen is involved, the force will be sufficient to also carry upward a suspended gondola containing instruments or even men. The first such "balloons" were launched in France in 1783.

When there is relative motion between a solid and a gas there is friction, as there is between a solid and a liquid—though again the effect is much smaller where a gas rather than a liquid is involved. The friction with gas ("air resistance") is enough, however, to slow the velocity of projectiles to the point where it must be allowed for if an artilleryman is to aim correctly. Air resistance also prevents a complete and perfect interchange of kinetic and potential energy by dissipating some of the energy as heat (see page 97).

The downward force of the gravitational field is proportional to the total mass of the body, while the upward force of air resistance is proportional to the area of contact of the moving body with air in the direction of its motion. For compact and relatively heavy bodies, such as stones, bricks and lumps of metal, the gravitational force is high, while the contact with air is over a relatively limited area so that air resistance is low. In such cases motion is close enough to what it would be in a vacuum for Galileo to have been able to draw correct conclusions from his experiments.

For light bodies, the gravitational force is relatively low. If such bodies are also thin and flat (as leaves or feathers are, for instance), they present a relatively large area to the air, and air resistance is relatively high. In such cases, air resistance almost balances the gravitational force, and these light bodies therefore fall slowly (they would fall quickly in a vacuum); this slow rate of fall fooled the ancient Greek observers into believing there was an intrinsic connection between weight and the rate of free fall.

There is a terminal velocity reached in motion through air under the influence of a constant force such as gravity, since air resistance does not remain constant but increases with the velocity of an object through air. As velocity increases, air resistance eventually balances the gravitational force. For heavy, compact objects this terminal velocity is very high, but for light, flat objects it is quite low. Snowflakes quickly reach their low terminal velocity and accelerate no further though they fall for miles. If a compact object is suspended from a light flat one, the two objects together reach a far lower terminal velocity than the compact object would by itself, and this is why a parachute makes it possible to fall safely from great heights.

Again, there is a Bernoulli effect for gases as well as for liquids, and air pressure drops as the velocity of moving air increases. A jet of air moving across an orifice covers that orifice with a low-pressure area (or a "partial vacuum"). If a tube connected with the orifice dips into a liquid under normal atmospheric pressure, that liquid is pushed up the tube and is blown out in a fine spray.

When a baseball or a golf ball spins in the air, one side spins with the motion of the air flowing past the ball as it moves; the other side spins against the motion. The side that spins against

Bernoulli effect in gases

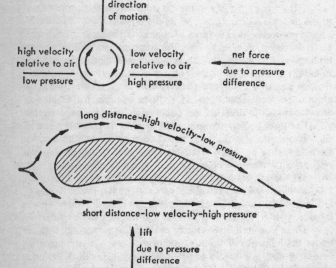

direction
of motion

high velocity
relative to air   low velocity
relative to air   net force

low pressure   high pressure   due to pressure
                                            difference

long distance–high velocity–low pressure

short distance–low velocity–high pressure

lift
due to pressure
difference

the motion has a greater velocity relative to the air, and the air pressure is less in that direction. The ball is pushed in the direction of lower pressure, so that the baseball curves in its flight (usually desirable, if it is the pitcher throwing the ball), while the golf ball "hooks" or "slices" (usually undesirable).

Where high velocity through air must be maintained with a minimum of force, streamlining is important. This importance increases with velocity, since air resistance also does. Thus, a horse and wagon need display no streamlining and automobiles need display very little. (A trend toward extreme automobile streamlining initiated in the late 1930's was a matter of appearance rather than necessity, and was abandoned.)

Airplanes, however, must be streamlined, and in reaching supersonic speeds it was not so much higher power that had to be developed, but the proper design for minimizing air resistance. Furthermore, airplane wings (themselves streamlined) are so designed that the air must move over a greater distance above than below, and so must move more rapidly above than below. This, by Bernoulli's principle, means there is less pressure above than below, and therefore a net upward force ("lift") that helps support the plane.

Boyle's Law

The properties of gases are of crucial importance with respect to the possibly atomic nature of matter. If matter is nonatomic, then variations in density must be caused by the intrinsic differences in the density of matter itself. Every bit of it, however small, must be as dense as every other bit. There would be no holes or empty space in matter, as there would be if the matter consisted of atoms.

If matter consisted of atoms, there might be space between the atoms, a space containing only vacuum. Matter might be made less dense, then, by pulling the atoms apart in some fashion so as to increase the proportion of empty space within a given volume. Conversely, matter might be made more dense by pushing the atoms together to reduce the proportion of empty space.

Indeed, it might seem that the density of a particular substance could be changed in just this fashion by heating or cooling. Density usually decreases with heating and increases with cooling. Thus, although the density of cold water is 1 gm/cm^3, that of hot water is only about 0.96 gm/cm^3.

Then again, solids melt to liquids if heated sufficiently, and the liquids become solids again if cooled. This change in the *state of matter* is accompanied by a sudden change in density. Thus, ice has a density of 0.92 gm/cm³, but as soon as it is melted to water the density increases sharply to 1.00 gm/cm³. Again, solid iron has a density of 7.8 gm/cm³, but this decreases sharply when iron is melted to the liquid form, which has a density of only 6.9 gm/cm³. An atomist might point out that a ready explanation for this is that in one state the constituent atoms are more compact than in the other. (Usually it is the solid state which is denser, with water a rather unusual exception.)

However, in all such changes, the density varies by only a few percent, and this is not overwhelmingly convincing. Working against the atomist is the fact that liquids and solids are relatively incompressible. Large increases of pressure (attainable only with specialized equipment) are required to bring about even small decreases in volume. For this reason, there can't be much empty space in ordinary matter, and even atomists must admit that in liquids and solids, atoms, if they exist, are in virtual contact. Since liquids and solids remain incompressible at any temperature, the feeling that atoms are further apart in hot water than in cold, or in liquid iron than in solid iron, seems to be wrong. If it were not wrong, then hot water and liquid iron would be at least moderately compressible, and they are not.

It is another matter entirely, however, when the point of view shifts from solids and liquids to gases. When liquid water is boiled and gaseous steam is formed, the change in density is drastic and dramatic. Where water has a density of 0.96 gm/cm³ at the boiling point, steam at the same temperature has a density of no more than 0.0006 gm/cm³. Steam is only about 1/1700 as dense as water.

This can be reasonably explained by adopting an atomistic view. One can suppose that the constituent atoms (or groups of atoms) making up water move far apart in the conversion of the liquid water to the gaseous steam, and that steam is as low in density as it is because it consists mostly of the empty space between atoms. We might generalize and say that whereas in liquids and solids atoms are virtually in contact, in gases they are far apart. This spreading out of atomic particles would account not only for the extremely low density of gases, but also for their low pressure, their small frictional forces, and so on.

If this atomistic view is so, and if the particles of gas are widely spread out, then gases ought to be easily compressible. If pressure is exerted upon a given volume of gas, that volume ought to decrease considerably. This is actually so, and the fact was first clearly presented to the scientific community by the English physicist Robert Boyle (1627–1691) in 1660.

He poured mercury into the open long end of a J-shaped tube and trapped some of the air in the closed short end. By adding additional mercury, he raised the pressure on the trapped air by an amount he could measure through the difference between the inches of mercury in the open and closed sides. He found that doubling the pressure on the trapped gas generally halved its volume; tripling the pressure reduced the volume to a third, and so on.

The trapped gas was always able to support the column of mercury on the other side once its volume had been reduced by the appropriate amount, so that the pressure it exerted was equal to the pressure exerted upon it. (This is to be expected from Newton's third law—which, however, was not yet enunciated in Boyle's time.)

Consequently, we can say that for a given quantity of gas the pressure (P) is inversely related to the volume (V), so as one goes up, the other goes down ($P = k/V$). Therefore, the product of the two remains constant:

$$PV = k \qquad \text{(Equation 10–1)}$$

This relationship is called *Boyle's law*.*

Another way of stating Boyle's law is as follows. Suppose that you have a sample of gas with a pressure P_1 and a volume V_1. If you change the pressure, either increasing it or decreasing it, to P_2, you will find that the volume automatically changes to V_2. However, the product of pressure and volume must remain constant by Equation 10–1, so we can say that for a given quantity of gas:

$$P_1 V_1 = P_2 V_2 \qquad \text{(Equation 10–2)}$$

and that, too, is an expression of Boyle's law.

Indeed, gas is so easily compressed that the pressure of the upper layers of a column of gas will compress the lower layers. Whereas a column of virtually incompressible liquid has a constant density throughout, columns of gases vary considerably in density with height. This is particularly noticeable in the case of the atmosphere itself.

If gas were as incompressible as liquid, and if it were as dense at all heights as it is at sea level, then one could easily calculate

---

* Boyle's law, as it turned out, is only an approximation (see page 208), but it is a very useful approximation and, in the case of some gases, an approximation very close to the truth.

what the height of the atmosphere ought to be. Air pressure is 1033.2 grams of weight per square centimeter. This means that a column of gas one square centimeter in cross section and extending straight upward to the top of the atmosphere weighs 1033.2 grams. A column with such a cross section, but only one centimeter high, has a weight of 1.3 milligrams. Each additional height of one centimeter added to the column would add an additional 1.3 milligrams, and it would take a total height of about 800,000 centimeters to account for the 1033.2 grams of air pressure. This is a height of just about five miles.

However, this cannot be right, for balloons have found air to exist at heights of over 20 miles, and less direct methods of measurement have shown perceptible quantities of air to exist at heights of over 100 miles.

The point is that the atmosphere is not at constant density. As one moves upward, one finds that a given quantity of gas is under less pressure because the quantity of air above it has become less. By Boyle's law, that given quantity of gas must therefore take up a larger volume. Consequently, as one rises, the amount of atmosphere remaining above, while decreasing rapidly in weight, decreases only very slowly in volume. For that reason, indeed, the atmosphere has no definite upper edge, but fades slowly off for hundreds of miles above earth's surface, decreasing in density until it peters out into the incredibly thin wisps of gas that make up interplanetary space.

By pointing out the atomistic argument first, I was trying to make absolutely clear the importance and significance of Boyle's experiment. It is not to be supposed, however, that that one experiment at once turned scientific opinion toward atomism. It was not until the first decades of the nineteenth century, a century and a half after Boyle's experiment, that the weight of evidence had finally accumulated to the point where scientists could no longer avoid accepting atomism.

The scientist usually given credit for the final establishment of atomism is the English chemist John Dalton (1776–1844). He worked out the "modern atomic theory" in detail, between 1803 and 1808, basing it chiefly on the observations of the properties of gases that had begun with Boyle's experiments. (In fact, one might maintain that Boyle's law made atomism inevitable, and that all that followed merely served to place a finer edge on the concept.)

It is now generally accepted that all matter consists of *atoms*; that these atoms may exist singly, but much more commonly exist in groups of from two to many hundreds of thousands; and that these groups of atoms, called *molecules*, maintain their identity under ordinary circumstances and form the particles of matter.*

It was by considering gases to consist of a collection of widely-spaced molecules (or, occasionally, of widely-spaced individual atoms) that it became possible to view such phenomena as sound and heat in a new and more fundamental manner.

* Under certain circumstances, molecules do alter in nature, and old combinations of atoms shift and change into new combinations. These shifts and changes of molecular combinations are the prime concern of the science of chemistry.

CHAPTER **11**

Sound

Water Waves

Fluids can move in the various fashions that solids can move. They can undergo translational motion, as when rivers flow or winds blow. They can undergo rotational motion, as in whirlpools and tornadoes. Finally, they can undergo vibrational motion. It is the last that concerns us now, for a vibration can produce a distortion in shape that will travel outward. Such a moving shape-distortion is called a *wave*. While waves are produced in solids, they are most clearly visible and noticeable on the surface of a liquid. It is in connection with water surfaces, indeed, that early man first grew aware of waves.

If a stone is dropped into the middle of a quiet stretch of water, the weight of the stone pushes down on the water with which it comes into contact and a depression is created. Water is virtually incompressible, so room must be made for the water that is pushed downward. This can only be done by raising the water in the immediate neighborhood of the fallen stone, so the central depression is surrounded by a ring of raised water.

The ring of raised water falls back under the pull of gravity, and its weight acts like the original weight of the stone. It pushes the water underneath downward and throws up a wider ring of water a bit farther away from the original center of disturbance.

148

This continues, and the ring of upraised water moves farther and farther out from the center. As it moves outward, the total mass of upraised water must be spread out through a larger and larger circumference, and the height of the upraised ring is therefore lower and lower.

Nor is there a single wave emanating from the center of the disturbance. As the initial wall of upraised water immediately about the center of disturbance comes down, it not only pushes up a wall of water beyond itself, but also pushes up the water at the center. This rises and then drops again, acting, so to speak, like a second stone, and setting up a second circular wall of water that spreads outward inside the first wall. This is followed by a third wall, and so on. Each successive wall is lower than the one before, since with each rise and fall of water some of the energy is consumed in overcoming the internal friction of the water and is converted into heat. As a particular wall of water spreads outward, some of its energy is also being continually converted into heat. Eventually, all the waves die out and the pool is quiet again; however, it is very slightly warmer for having absorbed the kinetic energy of the falling stone.

To produce a wave, then, we need an initial disturbance. If this initial disturbance, in correcting itself, disturbs a neighboring region in a fashion similar to the original disturbance, the wave is propagated.

In a propagated wave, if we concentrate our attention on a given point in space, we see that some property waxes and wanes, often periodically. In the case of water waves, for instance, if we view one portion of the water surface and no other, then the varying property is potential energy as that portion of the surface first rises, then falls, then rises again.

It is important to realize that the water is moving up and down only. The disturbance is propagated outward across the surface of the water, and it appears to the casual observer that water is moving outward; however, it is not! Only the disturbance is. A chip of wood floating on water that has been disturbed into ripples will rise and fall with the rise and fall of the water it rests on, but the moving ripples will not carry the wood with it. (To be sure, waves approaching the shore will carry material with them, sometimes even forcefully, as the water dashes on the rocks or beach. These waves, however, are being driven by the horizontal force of the wind and are different from the ripples set up by the vertical force of a falling stone.)

Suppose we imagine a cross section of the surface of the

water that is undergoing the disturbance of a falling stone. Ideally, ignoring loss of height with increasing circumference or loss of energy as heat, we have a steady rise and fall. This rise and fall is what we commonly think of when the word "wave" is spoken, or when we speak of a "wavy line."

In its simplest form, such a wavy line is identical with the type of curve produced if one plots the value of the sine of an angle (see page 111) on graph paper as the size of the angle increases steadily. For an angle of 0°, the sine is 0. As the angle increases, the sine also increases, first quickly, then more and more slowly, till it skims a maximum of 1 at 90°. For still larger angles it begins to decrease, first slowly, then more and more rapidly, reaching 0 again at 180° and passing into negative values thereafter. It skims a minimum of —1 at 270°, then increases again to reach 0 once more at 360°. An angle of 360° can be considered equivalent to one of 0°, so the whole process can be viewed as beginning again and continuing onward indefinitely. In plotting the graph, then, one gets a wave-like figure that can extend outward forever as it oscillates regularly between +1 and —1. It is this wave-like figure (the *sine curve*) that represents the shape of an idealized water wave.

A wave like the water wave, in which the motion of each part is in one direction (up-and-down in this case), and the direction of propagation of the disturbance is at right angles to that direction (outward across the water surface in this case), is a *transverse wave*. (Transverse is from Latin words meaning "lying across"; the motion of the water itself "lies across" the line of propagation.)

The point at which the disturbance is greatest in the upward direction (+1, in the sine curve) is the *crest*, and the point at which it is greatest in the downward direction (—1, in the sine curve) is the *trough*. Between crest and trough are points where the water is momentarily at the level it would be at if the surface were undisturbed (0, in the sine curve); these are *nodes*. There are two kinds of nodes in these water waves, for water may pass through a node on its way down to a trough or on its way up to a crest. We might distinguish these as "descending nodes" and "ascending nodes" (borrowing terms that are used in astronomy for an analogous purpose). The vertical distance from a node to either a crest or a trough is the *amplitude* of the wave.

Two or more points that occupy the same relative positions in the sine curve are said to be *in phase*. For instance, the points on the various crests are all in phase; so are the points on the various troughs. The ascending nodes are all in phase; the descending nodes are all in phase. All points lying a fixed portion of the way between an ascending node and a crest are in phase, and so on. If two waves exist, and if they match up in such a way that the crest of one is even in space or form with the crest of another at the same instant in time, those portions of the two waves are said to be in phase. It is possible that the entire stretch of both waves may be in phase in this fashion, crest for crest and trough for trough.

Naturally, points on a single wave that are not in phase are *out of phase*. And a pair of waves in which the crest in one case does not appear at the same time as the crest of another are out of phase.

A sine curve can be looked upon as consisting of a particular

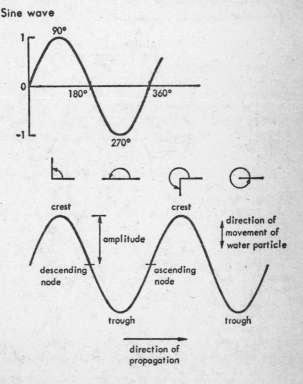

Sine wave

amplitude

direction of movement of water particle

crest

crest

descending node

ascending node

trough

trough

direction of propagation

small portion that repeats itself indefinitely. For instance, a portion of the sine curve from one crest to the next can be shaped into a stamp, if you like, and the entire sine curve can be reproduced by stamping that one crest-to-crest portion, then another like it to its right, another like it to the right again, and so on. The same could be done if we took a portion of the sine curve from trough to trough or from ascending node to ascending node or from descending node to descending node, and so on. An appropriate stamp can be made covering the section from any point on the

Wavelengths

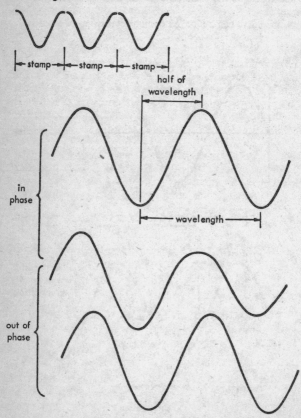

sine curve to the next point in phase. The length from any such point to the next is constant for any particular sine curve. This length (let us say from crest to crest for simplicity's sake) defines a *wavelength*. The wavelength is usually symbolized by the Greek letter "lambda" (λ).

If nodes are not distinguished from each other (and they are usually not in physics), then successive nodes are half a wavelength apart. If crest and trough are lumped together as *antinodes* (as sometimes they are), then successive antinodes are half a wavelength apart.

A particular crest moves outward along the surface of the water (though the water itself, I repeat, does not move outward with it), and the distance it travels in one second is the velocity of the wave.

Let us suppose that the velocity of a particular wave is ten meters per second and that the wavelength (that is, the distance from crest to crest) is two meters. If we fix our attention on a certain point of the water's surface, we will note that a particular crest is at that point. It travels outward, and two meters after it a second crest occupies that point; two meters after that there is is a third crest, and so on. After one second the original crest is ten meters away, and a fifth crest (10 divided by 2) is occupying the original spot.

The number of crests (or the number of troughs, ascending nodes, descending nodes, or any successive points in phase) passing a given point in one second is the *frequency* of the wave. Frequency is usually symbolized by the Greek letter "nu" (ν).

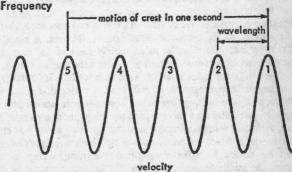

Frequency

$$\text{frequency} = \frac{\text{velocity}}{\text{wavelength}} = 5 \text{ per second}$$

From what I have said, it should be clear that the velocity of a wave divided by its wavelength is equal to its frequency, so:

$$v = \frac{\nu}{\lambda} \qquad \text{(Equation 11–1)}$$

The units of velocity are meters per second in the mks system, and those of wavelengths are meters. The units of frequency are therefore $(m/sec)/m$, or $1/sec$. Since in algebra $1/a$ is said to be the "reciprocal" of a, $1/sec$ is sometimes spoken of as *reciprocal seconds*. More often it is referred to by the phrase "per second." The frequency of the wave mentioned above as an example can be written $5/sec$, and this can be read as "5 per second" or as "5 reciprocal seconds."

Sound Waves

At a comparatively early stage in the quest for knowledge, sound came to be thought of as resulting from a kind of wave motion. The first experiments on sound were conducted by the ancient Greeks, and these were rather remarkable in one way, for the study of sound was one branch of physics in which the Greeks seemed, by modern criteria, to start off in the right direction from the very beginning.

As early as the sixth century B.C., Pythagoras of Samos was studying the sound produced by plucked strings. It could be seen that a string vibrated when plucked. The plucked string's motion was only a blur, but even so, certain facts about that blur could be associated with sound. The width of the blurred motion seemed to correspond to the loudness of the sound. As the vibration died down and the blur narrowed, the sound grew softer. And when the vibration stopped, either by natural slowing or by an abrupt touch of the hand, so did the sound. Furthermore, it could be made out that shorter strings vibrated more rapidly than longer ones, and the more rapid vibration seemed to produce the shriller sound.

By 400 B.C., Archytas of Tarentum (420?–360? B.C.), a member of the Pythagorean school, was suggesting that sound was produced by the striking together of bodies—swift motion producing high pitch and slow motion producing low pitch. By about 350 B.C., Aristotle was pointing out that the vibrating string was striking the air; and that the portion of the air which was struck must in turn be moved to strike a neighboring portion, which in turn struck the next portion, and so on. To Aristotle, then, it seemed that air was necessary as a medium through which sound was conducted, and he reasoned that sound would not be conducted through a vacuum. (In this, Aristotle was correct.)

Since in rapid rhythm a vibrating string strikes the air not once but many times, not one blow, but a long series of blows, must be conducted by the air. The Roman engineer Marcus Vitruvius Pollio, writing in the first century B.C., suggested that the air did not merely move, but vibrated, and that it did so in response to the vibrations of the string. It was these air vibrations, he held, that we heard as sound.

Finally, about 500 A.D., the Roman philosopher Anicius Manilius Severinus Boethius (480?–524?) made the specific comparison of the conduction of sound through the air with the waves produced in calm water by a dropping pebble. While this analogy has its value, and while water waves can be used to this day (and are so used in this book, for instance) to serve as a preliminary to a consideration of sound waves, there are nevertheless important differences between water waves and sound waves.

Transverse waves, such as water waves, can appear only under certain conditions. Such waves represent conditions in which one section of a body moves sideways with respect to another and then reverses that motion. (You can produce a transverse wave in a tall stack of cards by moving each card sideways by the proper amount.) Such a sideways motion is produced by a type of force called a *shear*. For such a force to result in a transverse wave, however, the force producing the shear must be countered by another force that brings the portions of the body back into line.

Within a solid, for instance, a blow may cause a portion of the substance to move sideways with respect to a neighboring portion. The strong cohesive forces between the molecules of a solid, which tend to keep each molecule in place, act to bring the displaced section back. It shoots back, overshoots the mark, shoots back again, overshoots the mark again, and so on. The resulting vibration is propagated just as the waves on a water surface are, and as a result it is possible to have transverse waves through the body of a solid.

The cohesive forces in liquids and gases are, however, very weak in comparison to those in solids and do not serve to restore a shear. If a portion of water or air is shifted sideways with respect to a neighboring portion, additional water or air will simply flow into the region left "empty" by the shifting portion, and the new arrangement of portions will remain. There are therefore no transverse waves through the body of a fluid.

To be sure, transverse waves will travel over the horizontal upper surface of liquids, for there we have the special case of an outside force, gravity, resisting the up-and-down shear. Within the body of the liquid, gravity cannot be counted upon to do this work, for each fragment of water is buoyed up by the surrounding water. Since the density of each bit of water is equal to the density of the surrounding water, each bit of water has a weight of zero (see page 124) and does not respond to gravity. If a portion of water within the body of the liquid is raised by a shear, it remains in the new position, in spite of gravity. Since transverse waves are confined to the surface of a fluid, and since gases have no definite volume and therefore no definite surface, it follows that transverse waves cannot be transmitted by gases under any condition.

Consequently, if sound is transmitted through the air as a wave form (as all the evidence indicates it must be), that wave form cannot be transverse. A logical alternative is that it consists of periodic compressions and rarefactions.

Consider the vibrations of a tuning fork, for instance. The prong of a tuning fork moves right, left, right, left, in a rapid periodic motion. As it moves right, the molecules of air lying immediately to the right are pushed together, forming a small volume of compression. The pressure within the compressed volume is greater than in the neighboring volume of normal air. The molecules in the compressed volume spring apart and push against the neighboring volume, compressing it. The neighboring volume compresses its neighbor as it springs apart, and so on. Thus, a volume of compression is propagated outward in all directions, forming an expanding sphere about the source of the disturbance just as the crest of a water wave forms an expanding circle about *its* source of disturbance. (The atmosphere is a three-dimensional medium, the surface of the water a two-dimensional one, which is why we have an expanding sphere in one case, an expanding circle in the other.)

Meanwhile, the prong of the tuning fork, having moved to the right and set off an expanding volume of compression, next moves to the left. More room is made to the right of the prong and the air in that immediate volume expands and becomes relatively rarefied. Pressure is higher in the neighboring un-rarefied air, which therefore pushes into the rarefied volume and is itself rarefied in the process. In this way, a volume of rarefaction expands outward on the heels of the volume of compression.

Again the prong of the tuning fork moves right, then left, then right, so that volumes of compression and volumes of rarefaction follow each other outward in rapid alternation for as long as the prong continues to vibrate. Each period of the prong (one movement back and forth) sets up one compression/rarefaction combination.

In these waves of alternate compression and rarefaction, the individual molecules of air move in one direction when compressed, then in the reverse direction when rarefied; the volumes of compression and rarefaction move outward and are propagated in a direction parallel to the back and forth motion of the molecules. Such a wave, in which the particles move parallel to the propagation rather than perpendicular to it, is a *longitudinal wave*, or a *compression wave*.

Longitudinal waves are harder to picture and grasp than are transverse waves, for there are no examples out of common experience that we can draw on to illustrate the former in the way we used water waves to illustrate the latter. Nevertheless, having gone into some detail about transverse waves, we can deal with longitudinal waves by analogy.

The points of maximum compression are analogous to the crests of transverse waves, and the points of maximum rarefaction to the troughs. In between there are areas where pressure is momentarily normal, and these correspond to the nodes.

The distance between points of maximum compression (or between points of maximum rarefaction) is the wavelength of the longitudinal wave. The number of points of maximum compression (or of maximum rarefaction) passing a given position in one second is the frequency of the longitudinal wave.

Since the molecules of liquids and solids, as well as those of gases, evolve a restoring counterforce when compressed,

Sound waves

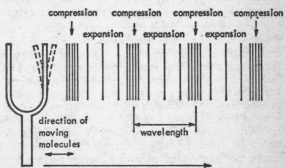

longitudinal waves can be carried through gases, liquids and solids. Sound waves in particular are carried by water and by steel, as well as by air. (Waves produced in the body of the earth by the vibrations induced by earthquakes are of both varieties, transverse and longitudinal. Both can be transmitted by the solid matter of the earth, but it was found that when the waves penetrated a certain depth below the earth's surface, only the longitudinal ones continued onward, while the transverse waves were stopped suddenly and entirely. It was from this that geologists were able to deduce that the earth contains a liquid core, and to measure its diameter with considerable accuracy.)

Sound waves, however, cannot be conducted in the complete absence of molecules. If an electric bell is suspended in a bell jar and set to ringing, it will be heard through the glass (which can carry sound waves). If the bell jar is gradually evacuated, the sound of the bell will become fainter and will eventually fade out altogether. The clapper may continue to strike the bell furiously and the bell may even be seen to vibrate, but no longitudinal waves can be set up among the molecules of an air that does not exist. As a result, sound will not be heard.

(It is frequently stated that the moon, which lacks an atmosphere, is a soundless world. However, sound can be transmitted through the moon's crust, and an astronaut may hear a distant explosion if he makes the proper contact with the moon's surface.)

Loudness

Suppose we consider a sound wave in which the succession of compressions and rarefactions are regular. This would be analogous to a transverse wave that had the form of a regular sine curve. Such a sound wave is heard by us as a steady musical note and is produced by a tuning fork. Indeed, if a pen is attached to the prong of a tuning fork in such a way that it makes contact with a roll of paper being moved at constant velocity in a direction at right angles to the vibration of the prong, a sine wave will be produced.

A tuning fork may produce sounds that differ in loudness. If it is struck lightly, it will emit a soft sound; if struck more heavily it will emit a sound which our ear will detect as identical with the first except for being louder. The lightly struck tuning fork will move back and forth over a comparatively small arc; the more heavily struck one will move back and forth over a larger arc. As is to be expected of simple harmonic motion, the two movements will involve an identical period despite the difference in amplitude, so either way the same number of volumes of compression and of rarefactions are set up per second. The frequency of the sound produced is, therefore, the same in either case.

However, the more heavily struck tuning fork, moving in the larger arc, compresses the air more violently. Therefore, a louder note differs from a softer note in that the compressed volumes of the former are more compressed, and the rarefied volumes more rarefied. The greater difference in extent of compression in a longitudinal wave is analogous to a greater amplitude in a transverse wave. This can easily be visualized if we think again of the tuning fork with the pen attached. A gently vibrating prong would mark off a sine curve of small amplitude; one vibrating through a greater arc, as a result of a heavier blow, would mark off a sine curve of greater amplitude.

To compress air against the resistance of its pressure requires energy, and the compressed air contains a store of energy that it can expend by expanding and pushing whatever is in the neighborhood. For this reason, sound waves can be considered a form of energy.

The more the air is compressed, the more energy it contains and can expend. Another way of looking at it is to consider that the vibrating tuning fork has kinetic energy that is expended in compressing air.* If the prong swings through a greater arc but completes its period in the same time, it moves at a greater average velocity and has more kinetic energy that it can expend in compressing air. Whichever way we look at it, we can come to the conclusion that loudness is a matter of quantity of energy, and that a loud sound contains more energy than a soft one.

The loudness, or *intensity*, of sound is measured in terms of the quantity of energy passing each second through one square centimeter of area, the area being perpendicular to the direction of propagation of the sound. Energy expended per unit time is power, and the amount of power involved in sound is very small. To indicate just how small, let's reconsider some units of power.

A watt is the mks unit of power and is equal to one joule per second. We are familiar with watts in connection with light

---

* The tuning fork, or any sound-producing device, also rarefies air against its own pressure, which also requires energy. The argument is exactly analogous if rarefaction is considered rather than compression.

bulbs; we all know that a light with a power of 75 watts is none too bright for reading purposes, and that one with as little power as 40 watts is rather dim. Even a night light, just bright enough to dispel the worst of the shadows and enable us to get to the bathroom at night without tripping over the furniture, has a power of 1/4 watt. A microwatt is 1/1,000,000 of a watt, so that such a night light has a power of 250,000 microwatts.

In comparison with that, ordinary conversational sounds carry a power of but 1000 microwatts, and low sounds sink down to bare fractions of a microwatt.

The ear detects differences in loudness by ratios of power rather than by actual differences. Thus, a 2000-microwatt sound will seem a certain amount louder than a 1000-microwatt sound, but a 3000-microwatt sound will not appear louder by as much again. It takes a 4000-microwatt sound to seem louder by as much as a 2000-microwatt sound is louder than a 1000-microwatt sound. To get a sound that is as much louder still than a 4000-microwatt sound, we must rise to an 8000-microwatt sound. The ratios 2000/1000, 4000/2000, and 8000/4000 are all equal even though the differences are not, and it is by ratios that the ear judges.

This means that the ear acts not by the power of a sound, but by the logarithm* of that power. When one sound carries ten times the power of a second sound, the ratio of the power of the first to that of the second is 10, and the logarithm of that ratio is 1. The difference in sound intensity is then said to be·one *bel*, so named in honor of Alexander Graham Bell (1847–1922), who studied the physics of sound and invented the telephone. Similarly, if one sound is 100 times as powerful as another, it is two bels

* The common logarithm of a number is its exponent when it is expressed as a power of 10. For instance, 10^2 is (10) (10), or 100; 10^3 is (10)(10)(10), or 1000. Therefore, the logarithm of 100 is 2 and that of 1000 is 3. The use of logarithms converts a geometric series (one in which each number is obtained by multiplying the preceding number by a fixed quantity) into an arithmetic one (where each number is obtained from the preceding by addition). In the series 10—100—1000—10,000—100,000, etc., each number is obtained by multiplying the previous number by 10. If the logarithms of the numbers in the series are written instead, we have 1—2—3—4—5, etc., where each number is obtained by adding 1 to the previous number. Our senses generally work by converting a geometric series to an arithmetic one in this fashion. If one stimulation is 100,000 times as intense as another of the same sort, the sense organ, working by logarithms, detects it as, say, five times as intense. In this way sense organs can be useful over an enormous range of intensity. This is the *Weber-Fechner Law*, so named in honor of two Germans: Ernst Heinrich Weber (1795–1878), who first expressed the law; and Gustav Theodor Fechner (1801–1887), who popularized it.

louder; if it is 1000 times as powerful it is three bels louder, and so on. This kind of unit imitates the logarithmic working of the ear.

The bel is rather too large a unit for convenience. A tenth of a bel is a *decibel*. One sound is a decibel louder than another sound when the first is 1.26 times as powerful as the second, for the logarithm of 1.26 is just about 0.1.

Because of the small amount of energy represented by even loud sounds, sound energy is not something we are usually aware of. The energy of a roll of thunder may be sufficient to cause objects to vibrate noticeably. The telephone is an example of the manner in which human ingenuity has managed to usefully convert sound energy into electrical energy and back to sound energy.

For the most part, however, the sounds that continually surround us, whether created by human beings, by other forms of life, or by inanimate surroundings, simply fade out and are converted into heat.

If sound remained unconverted into other forms of energy, we could easily see how the loudness of sound would fall off with distance from the source. The sound wave moves outward as an expanding sphere from the source, and the total power represented by each sound wave spreads out over that surface. The surface of a sphere is equal to $4\pi r^2$, where r is the radius of the sphere—that is, the distance from the source. If the distance from the center is tripled, the surface area is increased ninefold and only one-ninth as much power passes through any square centimeter on the surface. The intensity of sound would then be expected to vary inversely as the square of the distance from the source. This is how the intensity of gravitational attraction falls off, for instance. However, gravity is not absorbed by matter, whereas sound is easily absorbed by most of the objects with which it makes contact—even by the air itself. As a result, sound falls off more rapidly than one would expect.

CHAPTER 12

Pitch

The Velocity of Sound

A particular object has some natural period of vibration, and in the case of simple harmonic motion at least, this period is proportional to the square root of the mass of the object divided by the restoring force (see Equation 8–4 on page 107). In the case of the pendulum, where the restoring force is gravity (which increases with mass), the period varies as the square root of the length of the pendulum divided by the acceleration due to gravity (see Equation 8–9, page 112).

This means that we will generally expect that of two similar objects the larger and more massive will have the longer period of vibration. It will consequently produce fewer sound waves per unit time, and the individual waves will have a longer wavelength and a lower frequency.

The period of vibration can also be varied by changing the size of the restoring force, the period shortening as the restoring force increases in size. A taut string is more difficult to pull out of its equilibrium position than a slack one is, and from that it is clear that the force tending to restore the string to position is increased as the string grows tauter. Of two strings otherwise alike, the tauter snaps back faster and, if it is a bowstring, shoots the arrow farther. (That is why bowstrings are kept as taut as possible when the bow is in action.) A taut string, snapping back quickly, naturally has a shorter period of vibration than a slack one has and produces sound waves with higher frequency and shorter wavelength.

From experience, however, we know that all the factors that serve to produce a sound wave of low frequency also produce a deep tone, while those that bring about a sound wave of high frequency also produce a shrill tone. Large objects with long periods of vibration produce deep tones, while similar small objects produce shrill ones. Compare the tolling of a church bell with the tinkling of a sleigh bell, the strum of the string on the bass viol with the shrillness of the string on the tenor violin. In the realm of life, compare the trumpeting of the elephant with the squeak of the mouse; the honk of the goose with the tweet of the canary. The voice of a man with his longer vocal cords is deeper than those of women and children with their shorter ones. An individual can vary the shrillness of the sound he produces by adjusting the tautness of his vocal cords (though he is not aware he is doing so), and the sound of a freely vibrating string can be made more shrill as it is made more taut.

This property of shrillness, or depth in a tone, is referred to as the *pitch* of the sound, and it is quite obvious that the ear differentiates the frequencies of sound waves as pitch. As frequency increases, a sound is heard as increasingly shrill. As frequency decreases, a sound is heard as increasingly deep.

It is easy to determine the frequency of a sound wave. The vibrations of a tuning fork can actually be counted in several ways, including (to mention a simple method) having it mark itself by penpoint on a moving scroll of paper and counting the waves produced in a unit time. In this way, frequency and pitch can be matched. For instance, a tuning fork or pitch pipe that produces a "standard A" (the pitch against which musicians standardize their instruments) can be shown to have a frequency of 440 per second.

To calculate the actual wavelength of a sound of a certain pitch, one can make use of Equation 11–1. This tells us that the frequency (v) is equal to the velocity of the wave (v) divided by the wavelength (λ). Solving Equation 11–1 for λ, we find that:

$$\lambda = \frac{v}{v}$$
(Equation 12–1)

The piece of information we need to make Equation 12–1 useful is the velocity of sound. This velocity may be determined with considerable accuracy by a straightforward experiment which was first carried through successfully in the early seventeenth century.

Suppose a cannon is set up on one hill and observers are stationed on another hill a known distance away. When the cannon is fired, the flash is seen at once (assuming that light travels so quickly that its journey from one hill to the other takes up virtually zero time—which is correct). The sound of the cannon, however, is heard only after a measurable interval of time. The distance between cannon and observers divided by the number of seconds of lag in hearing the cannon (the possession of a good timepiece is assumed) will give the velocity of sound.

To be sure, if there is a wind the compression waves will be hastened onward by the overall movement of the air, or slowed down, depending on the direction of the wind. What can be done, therefore, is to place cannon on both hills and fire each, first one then the other. Whatever effect the wind has in one direction, it has a precisely opposite effect in the other, and averaging the two velocities obtained will give the velocity in quiet air.

The currently accepted velocity of sound at ordinary temperatures (say, 20° C or, what is equivalent, 68° F)* is 344 meters per second (or 1130 feet per second, or 758 miles per hour). This velocity varies a bit with temperature. On a cold winter day, it may be as low as 330 meters per second; on a hot summer day, as high as 355 meters per second.

The temperature difference has important effects. During the day, upper levels of the atmosphere are generally cooler than air at ground level. As the upper part of a beam of sound waves penetrates the cooler strata, it slows up; the effect of that is to veer the entire beam upward. (If you are walking, and someone seizes your left arm, slowing that part of your body, you automatically veer leftward.) At night, the situation is reversed, for the upper levels are warmer than the lower levels. The upper part of a beam of sound waves will quicken, and the whole beam will veer downward. It is for this reason that sound can usually be heard more clearly and over greater distances by night than by day.

However, if we confine ourselves to room temperature, we may write Equation 12–1 as:

$$\lambda = \frac{344}{v} \qquad \text{(Equation 12–2)}$$

* The question of temperature and temperature scales will be taken up later in some detail (see page 181).

Until recent times, sound traveled at a velocity much greater than that of any man-made vehicle, so for practical purposes the velocity of sound did not concern the traveler. With the invention of the airplane, however, and with the steady increase in the velocities of which it was capable, the velocity of sound became of importance for reasons other than those involving the speed of communication.

It is the speed of the natural rebound of molecules after compression that dictates the rate at which a compressed area restores itself to normal and compresses the next area; so it is this speed of rebound that determines the velocity of sound. It is also the speed of the natural rebound of molecules after striking a speeding plane that makes it possible for air to "get out of the way" of the plane. As the plane approaches the velocity of sound, then, it approaches the velocity with which the air molecules can rebound. The plane begins to "chase after" the rebounding air molecules and, with increasing speed, more and more nearly catches them. Such a plane compresses the air ahead permanently (or at least for as long as it maintains its speed), since the air cannot get out of its way. This volume of compressed air ahead of the plane puts great strains upon the plane's structure; for a time in the 1940's, it was felt that a plane would disintegrate if it approached the speed of sound too closely. Thus, talk began to be heard of a "sound barrier," as though the velocity of sound represented a wall the plane could not break through.

The ratio of the velocity of an object to the velocity of sound in the medium in which the object is traveling is called the *Mach number*, in honor of an Austrian physicist, Ernst Mach (1838–1916), who toward the end of the nineteenth century first investigated the theoretical consequence of motion at such velocities. To equal the velocity of sound is to be moving at "Mach 1," to double it is to be at "Mach 2," and so on. A Mach number does not represent a definite velocity, but depends upon the nature, temperature, and density of the fluid through which the object is traveling. For normal air at room temperature, Mach 1 is 344 meters per second, or 758 miles per hour.

Improved design of planes enabled them to withstand the stresses at high velocities, and on October 14, 1947, a manned plane "broke the sound barrier" by traveling at a velocity of more than Mach 1. Since then velocities of Mach 3 and more have been attained. (An astronaut circling the earth at a speed of five miles per second might be said to be traveling at Mach 25, if the velocity of sound in ordinary air is used as a comparison. However, the astronaut is traveling through a near vacuum across which no significant amount of sound is conducted, and Mach numbers do not really apply to him.)

A plane traveling at *supersonic velocities* (velocities greater than Mach 1) carries its sound waves ahead of it, so to speak, since it travels more quickly than they could alone. The volumes of compression are brought together, and instead of a smooth progression from compression to rarefaction and back, as in ordinary sound waves, there is a sharp dividing line between a volume of strong compressions and the normal surrounding atmosphere. The strong compression streams backward in a cone-shaped band, with an angle depending on the Mach number, and is called a *shock wave*. A similar shock wave streams back from speeding bullets, too; it is also formed by the effect of lightning bolts, for instance, which will energetically expand air at velocities greater than Mach 1. (The shock wave is an example of a wave form that is not periodic.)

If a plane traveling at supersonic velocities slows down or veers off, the shock wave will revert to ordinary sound waves, carrying volumes of unusually strong compression and rarefaction, however. In this train, sound waves expand and weaken as they travel, but if they are fairly close to the ground to begin with, and happen to be directed downward, they will strike the ground with considerable strength, producing the now well-known "sonic boom."

Thunder is the sonic boom produced by lightning, and the crack of a bullwhip is a miniature sonic boom, since it has been established that the tip of such a whip can be made to travel at supersonic velocities.

The velocity of sound, when spoken of simply as such, always implies its velocity through air. However, sound travels through any material body, and its velocity varies with the nature of the body. Intermolecular forces in liquids and solids, stronger than in gases, bring about a much quicker rebound after compression. Consequently, sound travels with greater velocities through liquids and solids than through any gas, and the more rigid the substance (and hence the stronger the intermolecular forces), the greater the velocity of sound through it. In water, sound travels at a velocity of 1450 m/sec (3240 miles per hour), and in steel it travels at a velocity of about 5000 m/sec (or 11,200 miles per hour).

The Musical Scale

Sounds of different pitch can be produced in musical instruments by striking or plucking strings of different length and thickness, as in the piano or harp; by using few strings but altering their effective length by pinning one end with the finger at varying points, as in the case of the violin; by allowing sound waves to fill tubes which may be lengthened or shortened by physical movement, as in the trombone; or by blocking or unblocking certain sections of the tube by stopping a hole with a finger, as in a flute, or depressing a key, as in a trumpet.

When two notes are sounded, either together or one after the other, the combination is sometimes pleasant and sometimes unpleasant. This is partly a subjective and cultural matter, for we like what we are used to and many types of music, such as rock and roll or traditional Japanese, sound unpleasant to the uninitiated but very pleasant to the devotee. Nevertheless, if we confine ourselves to the serious music of the West, we can come to certain conclusions about this.

When two notes are sounded together, the result is not two separate trains of sound waves, each traveling independently through the air. Instead, the two waves add to each other to form a resultant wave.

To make things very simple, suppose that two sound waves are each of the same frequency but are sounded in such a way that one is half a wavelength behind the other. Whenever one sound wave is forming an area of compression at one point, the other is forming an area of rarefaction there, and vice versa. The two effects cancel each other and the air does not move. As a result, the two sounds taken together produce silence, and this phenomenon is called *interference*. It is difficult to picture this if we think of longitudinal waves. However, if the longitudinal waves are pictured as analogous transverse waves (as they invariably are for this purpose) interference is easily pictured. Wherever the sine curve of one sound wave goes up, the sine curve of the other goes down, and if the two are added together a horizontal line (no wave at all) is the result.

On the other hand, if two waves of the same frequency are sounded exactly in phase, they would add to each other, so compressed areas are more compressed and rarefied areas are more rarefied than if either sound had been produced alone In transverse wave analogy, the crests and troughs of the separate waves would match, and the resulting crests would be higher and troughs deeper of the two waves together than of either alone. The ear would hear one sound of the proper pitch, but louder. This is *reinforcement*.

Actually, perfect interference or reinforcement is unlikely. Instead, two or more waves will combine, reinforcing here, interfering there, and will form resulting patterns of very complicated form that will not at all resemble the regular sine waves of individual notes. However complicated these patterns may be, they will remain periodic. That is, a small unit section of the pattern can be taken, and the entire pattern can be shown to be made up of a succession of these units.

In 1807, the French physicist Jean Baptiste Joseph Fourier (1768–1830), studying wave forms generally, showed that any periodic wave pattern, however complicated it might seem, could be separated by appropriate mathematical techniques into the individual sine waves making it up. The mathematics involved is referred to as *harmonic analysis* because it can be applied to musical sounds. (The wave patterns of musical sounds are composed of separate sine waves that display an orderly set of interrelationships. Where this is not true, but where the component sine waves are chosen and combined at random, so to speak, the result is not

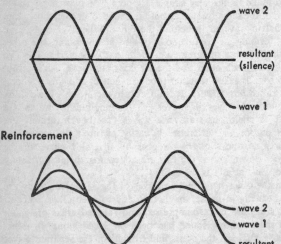

Interference

wave 2

resultant
(silence)

wave 1

Reinforcement

wave 2

wave 1

resultant

music, but "noise." The difference is analogous to that between an orderly but complicated geometrical figure and the same lines combined in random fashion to produce a scribble. Fourier's methods could be used to analyze the wave patterns of noise, too, so perhaps it should be referred to by the more neutral term *wave analysis*.)

Restricting ourselves to very simple cases, and without inserting appreciable mathematics, let's consider two notes of different pitch, and therefore of different frequency, sounded together. The compressed regions of the sound wave (or the crests, if we wish to speak in the more easily visualized transverse-wave analogy) would be coming at shorter intervals in the case of the note of higher frequency, and they would overtake those of the sound wave with the lower frequency.

Suppose one note has a frequency of 250 per second and another note of 251 per second. and suppose they start in phase. The first crest for both appears simultaneously. The second crest of the 251/sec note appears just a little sooner than the second crest of the 250/sec note. The third crest appears still sooner, and the fourth crest appears still sooner. At the end of one second, however, one note has completed exactly 250 vibrations and the other exactly 251 vibrations. They are back in phase, but the 251/sec note has gained one complete crest.* Each succeeding second, the 251/sec note gains another complete crest.

At the point where the two notes are in phase, crest for crest, there is a short period of complete reinforcement, and the note sounds loudly. As the second progresses and the crests fall more and more out of phase, there is more and more interference and the sound becomes softer. At the half-minute mark, midway between two in-phase periods, the notes are completely out of phase and the crests of one match the troughs of another, and there is a short period of complete interference. The result is a regular swelling and dying of sound, with the maximum loudness coming at second-intervals when the crests match. Such periodically changing loudness when two notes are sounded together is called a beat.

Suppose the two notes had frequencies of 250/sec and 252/sec, respectively. Then, after half a second, one note would have completed 125 vibrations and the other 126 vibrations, and

---

* The two notes are racing only in connection with the number of crests being produced in a given time, not in terms of velocity. Both notes are traveling through space at the same velocity. Indeed, the velocity of sound does not depend on frequency.

they would be back in phase with crest matching crest. This would be repeated every half-second, and there would be two beats per second. The number of beats per second, where two notes are sounded simultaneously, is generally equal to the difference in the frequencies of those notes.

If beats are infrequent enough to be heard separately, they render the sound combinations unpleasant to the ear. Apparently, 30 beats per second is maximally unpleasant. Where beats are more than 60 per second, however, they melt into each other as far as the ear is concerned, and the combination of sounds seems pleasant or harmonious.

Consider two notes of which one has a frequency exactly double the other. One has a frequency of 220/sec, let us say, and the other 440/sec; the ratio of frequencies is 1:2. The number of beats, when the notes are sounded together, is 440-220, or 220 a second. The beats duplicate the lower note, so the two notes seem to melt into each other and be almost the same note. They go well together.

It was Pythagoras who first noticed that notes that go well together are related by these small whole-number ratios. He had no method of measuring frequency itself, but he considered strings of different lengths. He found that two strings with lengths in a 1:2 ratio produced a pleasant combination; so did strings with a 2:3 ratio and a 3:4 ratio.

(Pythagoras wandered off in mystical fashion from these sound observations—sound in both senses. He assumed that the interplay of small whole numbers in the production of pleasing sounds fit in with his views that all the universe was ruled by number. He and his pupils speculated that the planets themselves produced sounds—the so-called music of the spheres—with notes based on their relative distances from the earth. Science did not free itself of these notions for 2000 years.)

Suppose then that we start with a note of a 440/sec frequency (the standard frequency for musicians) and call it A. A note of twice the frequency sounds so much like it that we can call that A, too, and we can use the letter for a sound of half the frequency, for that matter. In fact, we can have a whole series of such A's, with frequencies of 110/sec, 220/sec, 440/sec, 880/sec, 1760/sec, and so on, extending the range, if we choose, both upward and downward indefinitely.

Between any two successive A's, we can introduce other notes with frequencies that bear some orderly arithmetical relationship to the A-notes and to each other. It is customary to introduce six other notes in the interval; these are lettered B, C, D, E, F, and G. Thus we have, from A to A, the notes: A, B, C, D, E, F, G, A. In passing from A to A, there are eight notes (counting both A's) and seven intervals between notes. The span from A to A is therefore called an *octave* (from a Latin word for "eighth." Other spans are spoken of in plain English. The span from C to G (C, D, E, F, G), involving as it does five notes, is a *fifth,* while the span from C to F is a *fourth.*

The frequencies associated with the notes from the 220/sec A to the 880/sec A are:

A = 220	A = 440	A = 880
B = 247.5	B = 495	
C = 264	C = 528	
D = 297	D = 594	
E = 330	E = 660	
F = 352	F = 704	
G = 396	G = 792	

The range from 220/sec to 440/sec is one octave and that from 440/sec to 880/sec is another octave. Each note in the upper octave is double the corresponding note in the lower octave, so the interval from B to B is an octave; so is the interval from C to C, from D to D, and so on. Remembering to double the frequency for each higher octave and halve it for each lower octave, you can write the frequencies for any note in any octave.

If the successive notes within any octave are sounded, they sound just like the corresponding notes within any corresponding octave lower or higher. The standard piano keyboard covers a range of a little over seven octaves; if the white notes are sounded one after the other, one can easily detect the same "tune" to be repeated seven times, at successively higher pitches.

The frequencies are interrelated by ratios that can be expressed in small whole numbers. The ratio of G to C, for instance, is 396:264, or 3:2; the ratio of F to C is 352:264, or 4:3. It is these simple ratios that Pythagoras studied, and it is the simplicity of the ratios that sets up beats that reinforce the notes themselves

Pattern of octave intervals

and make them blend well together. That is why fifths and fourths are much used as intervals between successive notes.

Then, too, the ratios of C, E and G are 264:330:396, or 4:5:6, and the three notes sounded together as a *major triad* make a pleasing sound-combination, or *chord*. F, A and C also make a major triad, and so do G, B and D. In fact, the note intervals are so designed that every note can be part of one or another of these three major triads.

If the ratio of the frequency of adjacent notes is considered, it turns out that B:A as 9:8. The ratio for D:C and for G:F is also 9:8. The ratio for E:D and A:G is not quite that, but it is close, 10:9. In other words, of the seven intervals between the notes of the octave, five are of roughly equal size, and we can call them "whole intervals."

The frequency ratio of F:E, however, is only half as large, for it is 352:330, or 16:15; this is also true of the ratio of C:B. (This may be easier to see if we express it another way. A ratio of 9:8 represents an increase in frequency of 12.5 percent, and one of 10:9 represents an increase of 11.1 percent. The ratio 16:15, however, represents an increase of only 6.7 percent.) In passing from B to C or from E to F, then, we are traversing only a "half interval."

If we start from A and go up the notes through B, C, and so on, we will be passing intervals in the following pattern: whole, half, whole, whole, half, whole, whole, whole, half, whole, whole, half, and so on. Successive half intervals are separated by two whole intervals, three whole intervals, two whole intervals, three whole intervals, and so on.

When we sing the scale, using the traditional names for the notes (*do, re, mi, fa, sol, la, ti, do*), through long habit, we insist on placing the half-note intervals between *mi* and *fa* and between *ti* and *do*. Any other arrangement sounds wrong to us. We therefore want the seven intervals of the octave to fall into the following pattern: whole, whole, half, whole, whole, whole, half. If you check back, you will see that this particular arrangement can only be brought about if we start with *do* on the note C (it doesn't matter which C). Then *re* becomes D, *mi* becomes E, *fa* becomes F, *sol* becomes G, *la* becomes A, *ti* becomes B, and *do* is C again. The *mi-fa* half interval corresponds to the EF half interval, and the *ti-do* half interval corresponds to the BC half-note interval. The arrangement of notes you sound in singing the scale now matches the successive notes you tap out, beginning at C on the white keys of the piano. If you start on any white key of the piano other than

C and play the successive white keys, the piano and you will sound half intervals at different points in the scale, and the piano (not you, of course) will sound dreadful.

It is desirable to be able to play the scale from any point on the piano keyboard, so that the range of the scale can be adjusted to a particular human voice, for instance. For this reason, in every octave, five black notes are inserted to break up the five whole intervals. This accounts for the familiar black note pattern of two (CD, DE) and three (FG, GA, AB) all along the keyboard. Now the scale can be sounded by beginning at any note on the piano (either white or black) provided you remember to choose your notes carefully and play sometimes black and sometimes white. Only if you start on C, however, can you play the scale by sounding successive white notes only.

It is for this reason that C seems a natural *do* and that the "key of C" is the simplest key to play for beginners (white keys only, for the most part!). "Middle C" is the particular C that is about at the midpoint of the piano keyboard, and it is the C with the frequency 264/sec.*

Modification of Pitch

Pitch will change if the source of sound is moving relative to the hearer. Suppose a distant train, standing motionless, sounds a whistle that has a frequency of 344/sec. In that case, when the sound wave reaches us, 344 compression/rarefaction combinations will strike our eardrum each second. Since sound (at room temperature) travels at 344 m/sec, successive areas of compression are a meter apart.

Suppose next that the train is moving rapidly toward us at

* Physicists often use a frequency of 256/sec for middle C, because as a power of 2, 256 is a particularly easy number to halve and double. It is $(2)(2)(2)(2)(2)(2)(2)(2)$, or 2^8.

Piano keyboard

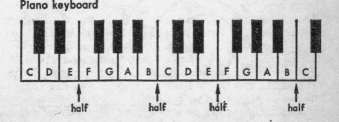

a rate of 34.4 m/sec (75.8 miles an hour), or just one-tenth the velocity of sound. It is still sounding its whistle. One region of compression is moving ahead of it, and by the time it has moved a meter, another region of compression is emitted. By that time, however, the train has moved forward a tenth of a meter and the second region of compression is only 0.9 meters behind the first. This happens for all successive regions of compression if the train maintains a steady pace. For this reason, sound waves from the whistle of the approaching train enter our eardrums 0.9 meters apart and in one second 344/0.9, or 382 of them, strike the eardrums. A person on the train, and therefore moving right along with the whistle, receives 344 regions of compression in one second. The ratio 382:344 is close to 9:8, so the sound is a whole interval shriller (see page 172) for the person watching the train approach than for the person on the train.

On the other hand, if the train were receding, then by the time a region of compression had moved a meter toward the hearer and a new region of compression was due, the train would have moved a tenth of a meter away, and the two areas of compression would be 1.1 meters apart. The frequency would be 344/1.1, or 312 per second. Now it is deeper by nearly a whole interval than it would sound to the person on the train.

If the train passed us at this velocity, the sound we heard would shift suddenly from a frequency of 382/sec as it was approaching and passed, to 312/sec as it passed and receded.

This phenomenon is called the Doppler effect in honor of the Austrian physicist Christian Johann Doppler (1803–1853), who first studied the effect and explained it correctly in 1842.

Pitch can be made to vary in a much more subtle way, too. The same note sounded with the same loudness on the piano, violin, and clarinet sounds different to us. If we have any experi-

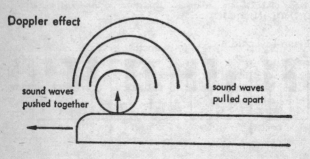

Doppler effect

sound waves pushed together

sound waves pulled apart

ence at all, we can tell which instrument is sounding the note. This difference in notes that are identical in pitch and loudness is a difference in *quality* or *timbre*.

To explain this, we must consider that the vibrations of a string, or of any sound-producing device, are actually more complicated than I have described them to be. A string, for instance, may indeed vibrate all in one piece to produce a vibration and, therefore, a sound wave of a given frequency. In the transverse-wave analogy, this would be a simple, regular sine curve, and is the *fundamental note*. It is the fundamental note we usually think of when we speak of the frequency of a particular note.

However, the string may also vibrate as two halves: one half moving to the right as the other half moves to the left and vice versa; the midpoint of the string, bounding the two halves, serving as a motionless *node*. Each half of the string vibrates at twice the frequency of the whole string, so a note is sounded with just twice the frequency of the fundamental note. The string may also vibrate in thirds, in fourths, in fifths, and so on, to produce notes with frequencies three times, four times, five times, and so on, that of the fundamental notes. All these notes of higher frequencies are called *overtones*. The fundamental note and the various overtones are sounding simultaneously; the actual motion of the string is a combination of all. The fundamental note remains dominant, but the overtones add their wave forms, and therefore the resulting wave form is far more complicated than a simple sine curve. Furthermore, for strings under different conditions (to say nothing of other sound-producing devices) the overtones may be receiving different proportionate stress; certain overtones may be stronger in some cases than in others, so the final wave form will be different for different instruments. Such a difference imposed on the eardrum is great enough for us to detect.

This difference can be magnified by methods of selecting some overtones from the rest for special magnifications. Let's see how.

A vibrating object may force another to vibrate in unison, so the second object sets up the same sound wave pattern and produces the same sound. If a vibrating tuning fork has its stem placed in contact with a table, its sound is suddenly louder because now the entire table is vibrating in unison.

Such a *forced vibration* need not even be the result of direct physical contact between solids. Indirect contact through air may be sufficient. A given vibration will set the air pulsing in longitudinal waves; these waves will, in turn, set the eardrum vibrating in unison. The eardrum will move inward when a region of compression strikes, and outward when a region of rarefaction does; it moves a greater distance from the equilibrium position as the regions become more compressed and more rarefied. It is through such forced vibrations that the eardrum exactly duplicates the original vibration, and we are able to judge (via a complicated hearing mechanism we will not describe here) the pitch, loudness, and even the timbre of a sound.

There are occasions, however, when one particular frequency is more easily "forced" than another on a second body. Imagine yourself pushing a child on a swing, for instance. The child on a swing represents a form of pendulum and has a natural period of vibration. If you apply successive pushes to the swing at random intervals, you will often push the swing as it is moving back toward you and will cancel what motion it possesses, slowing it. By persisting, you will keep the swing moving in accordance with your pushes, but you will expend a lot of energy doing so. If, however, you timed your pushes to match the natural period of vibration of the swing, you would push each time as the swing begins to move away from you, thus adding to its velocity and increasing it further with each swing and rhythmic push. At the expense of far less energy, you would get a far more rapid and extended swing.

(Marching soldiers crossing a bridge are supposed to break step. Otherwise, if the thud of the footsteps in unison happened to match the natural period of vibration of the bridge, the bridge would swing in wider and wider arcs until it finally broke apart.)

The situation is analogous for sound waves. The sound wave of a particular note would push another object with each region of compression and pull it with each region of rarefaction. If the rhythmic push-and-pull did not match the natural period of the receiving object, the forced vibration could only be obtained at the expense of considerable energy being used to overcome that natural period. If, however, the frequency of the note just matched the natural period of vibration of the receiving object, the latter would begin to vibrate more and more. This is called *resonance* (from Latin words meaning "to sound again").

Any given sound wave would produce far more vibration in a resonating object than in any other kind; in fact only the resonating object might produce sound waves strong enough to be audible. Suppose, for instance, you raise the top of a piano to expose the wires and step on the "loud" pedal to allow all those wires to vibrate freely. Now sing a short, loud note. Only those wires that vibrate at the frequency of that note will resonate, and when you stop singing, you will hear the piano answer back softly in that same note.

Musical instruments depend upon the resonance of the materials making up their structure to strengthen and add richness to the notes produced. Pianos have a "sounding board" just under the wires, and this device can resonate with the various notes. Without that board, the notes sounded by the wires would be quite weak.

Naturally, the resonating portions of each instrument, although resonating to almost all the notes (the resonating portions are complicated in shape and different parts have different natural periods of vibration), do not do so with equal efficiency. The wood of a violin resonates to the notes produced by it, but it may resonate more efficiently to some overtones than to others. No two violins are of exactly the same shape, or of exactly the same wood with the same grain arrangement, or possess exactly the same varnish. As a result, there are subtle differences in resonance from instrument to instrument. The Italian violinmaker Antonius Stradivarius (1644–1737) manufactured violins that are the despair of imitators, for it is almost impossible to duplicate their richness of tone.

The sounds we ourselves make produce resonances in the air filling the hollows in throats, mouths and nasal cavities. The natural vibrations of the air depend on the shape and size of the cavities, and since in no two individuals are these cavities of precisely the same shape and size, voices differ in quality; we usually have no trouble recognizing the voice of a friend from among a large number of others.

Reflection of Sound

A ripple in a water tank is turned back on itself when it strikes the rim of the tank; having progressed, let us say, leftward prior to contact, it proceeds rightward thereafter, much as a billiard ball does that has struck the edge of a pool table head-on. The water wave has been *reflected* (from Latin words meaning "to bend back").

Sound waves can be reflected, too. A mountain wall will reflect them, for instance. A word shouted across a valley is heard almost at once as it leaves the lips, being conducted through the air from lips to ear. It is then heard again seconds later, after the sound wave has reached the mountain wall, been reflected, and crossed the valley a second time. This is the *echo*. If mountain sides are properly arranged, more than one echo may be heard.

Similar echoes may be heard in tunnels, in large empty rooms, and indeed anywhere where hard surfaces reflect sound rather than absorb it. In ellipsoidal rooms a sound uttered at one focus of the ellipse will spread out in all directions; however, in reflecting from various portions of the walls and ceiling, it will concentrate upon the other focus. Two people standing at the foci can converse in whispers, even though separated by a large distance. Such "whispering galleries" always amaze those who have never encountered one before.

In rooms of moderate size, the length of time taken for a sound wave to travel to a wall, be reflected to an opposite wall, be reflected once again to the first wall, and so on, is so short that distinct echoes will not be heard. Instead, a series of very rapid echoes will blend into a dull, hollow rumble that may persist audibly for a considerable time after the original sound is no longer being formed. This persistence of sound is called *reverberation*. The study of the behavior of sound in enclosed places, particularly with regard to such reverberation, is called *acoustics* (from a Greek word meaning "to hear"), a term that is sometimes applied to the study of sound generally.

Reverberation can represent a great inconvenience. A lecturer may find that his words cannot be heard because of the dying sound of his previous words. An orchestra may find its best efforts reduced to discord as previous notes live on past the time when they are wanted or needed. Reverberation can be reduced by draping the walls, using a soft, pulpy material for the ceiling, or even by the presence of an audience in winter clothing. When sound waves enter the small interstices of fabric or other porous material, contact of the moving air molecules with solid material is made over a hugely increased area. Friction is increased and sound energy is converted into heat. The sound waves, in other words, are absorbed rather than reflected.

This can be overdone. If reverberation is reduced to too low a level, there seems a "deadness" to sounds. A reverberation period of one second, or even two if the room is very large, is aimed for.

Sound waves are not always either reflected or absorbed (nor are water waves). There is a third alternative: sound waves (and water waves, too) can bend around obstacles and continue onward. It is because of this that we have no difficulty hearing someone call from behind a tree or from around a corner. This ability to bend around obstacles is not the same for all kinds of waves.

In 1818, the French physicist Augustin Jean Fresnel (1788–1827) was able to show, in connection with his studies of wave motions generally, that whether a wave was reflected or not depended upon the comparative size of the wavelength and the obstacles. When an obstacle was the size of the wavelength or less, it did not reflect the wave, which, instead, bent round the obstacle. If the obstacle was considerably larger than the wavelength, the wave was reflected.

Consider the common sounds we hear about us every day, with frequencies, let us say, through the middle range of the piano from C below low C to C above high C—a range of four octaves. The range of frequencies extends from 66/sec to 1056/sec. If we make use of Equation 12–2, we see that the range of wavelengths over these four octaves is from 5.2 meters down to 0.32 meters (roughly from 1 to 18 feet, in common units). The common obstacles we meet with fall within this range of size and do not reflect such sound to any great degree, so sound bends.

This bending is, of course, more likely for the deeper sounds than for the shriller ones. We judge the direction of a sound by the inequality of loudness in the two ears, automatically turning our head until both ears hear the sound with equal loudness. Our head is large enough to reflect, somewhat, a shrill sound coming from one side. There is then a considerable reduction in the intensity of that sound making its way around our head to the other ear. We have no trouble, therefore, locating a child by its shrill cry. On the other hand, the deep tones of the lower register of an organ move around our head with ease and sound equally intense in both ears. The sound seems to come from all around us; this in itself lends majesty to the swell of the organ.

The full range of the piano covers 7.5 octaves. The lowest note possesses a frequency of 27.5/sec and a wavelength of 12.5 meters. We can hear still deeper sounds; however, the usual extreme in that direction is 15/sec, a sound with a wavelength of 22 meters. The highest note of the piano possesses a frequency of 4224/sec and a wavelength of 0.081 meters, or 8.1 centimeters. The adult human ear can hear sounds with a frequency as high as 15,000/sec (wavelength, 2.2 centimeters), and the child can sometimes hear a frequency as high as 20,000/sec (wavelength, 1.7 centimeters). Such extremely shrill sounds will be reflected quite well by objects too small to reflect sounds in the more common range. The high-pitched creak of a cricket may be reflected so well by various objects that it is next to impossible to tell exactly where the original sound is coming from.

It is, of course, possible for objects to vibrate with frequencies of less than 15/sec and more than 20,000/sec; when this happens sound waves are produced which are not audible. Those that are too deep to be audible are *infrasonic waves* (from Latin words meaning "below sound"), while those that are too shrill to be audible are *ultrasonic waves* (from Latin words meaning "beyond sound").

Infrasonic waves are comparatively unimportant except where they become energetic enough to do physical damage, as in earthquakes. Ultrasonic waves impinge upon us more often and in many ways. For one thing, they are not inaudible to all forms of life; many animals smaller than ourselves can both produce and hear them. The "silent" whistles to which dogs respond produce ultrasonic waves that they can hear though we cannot. The singing canary produces ultrasonic waves that would undoubtedly greatly add to the beauty of the song if we could but hear them. The squeaking mouse also produces them, and the waiting cat can hear them where we cannot—which increases the efficiency of the feline stalk.

Ultrasonic sound waves, with wavelengths even shorter than those of the shrillest sounds we can hear, can be reflected efficiently by quite small objects. Bats take advantage of this fact. They emit a continuous series of ultrasonic squeaks while flying. These have frequencies of from 40,000 to 80,000/sec and, therefore, have wavelengths of from 8 to 4 millimeters. A twig or an insect will tend to reflect such short wavelengths; and the bat, whose squeaks are of extremely short duration, will catch the faint echo between squeaks. It can thus guide its flight by hearing alone and continue flying with perfect efficiency even if blinded. This process is called *echolocation*.

Men duplicate this effect by making use of beams of ultrasonic waves underwater. These are reflected from objects such as the sea bottom, jutting rocks above the sea bottom itself, schools of fish, or submarines. The technique is referred to as *sonar*, an abbreviated form of "*so*und *n*avigation *a*nd *r*anging" (where "ranging" means getting the range of an object—that is, determining its distance).

CHAPTER **13**

Temperature

Hot and Cold

Heat has been mentioned several times in the book, notably toward the end of Chapter 7 in connection with the conservation of energy. I have not stopped, however, to consider it in detail, since to do so with proper understanding required first a consideration of the properties of fluids and, in particular, of gases. Enough of these properties have now been described to make it advisable to turn again to the subject of heat.

Heat is most familiar to us as a subjective sensation. We feel something to be "hot" or "cold," and we know what we mean when we say that one object is "hotter" than another. The degree of hotness or coldness of an object is called its *temperature*.

Temperature is of importance to physicists because a great many of the properties of matter with which he deals vary with temperature. In the previous chapter, for instance, I mentioned that the velocity of sound varied with temperature (see page 164). Again, the volume of a given mass of water increases as it is heated to near the boiling point, and so the density decreases. Hot water possesses weaker cohesive forces than does cold water, so the viscosity and surface tension decrease as temperature goes up. Even such seeming unchangeables as the length of an iron rod change with temperature

It follows then that if a physicist is to make proper generalizations concerning the universe, he must know just how the properties of matter change with temperature, and to do that he must be able to measure the temperature accurately. Our subjective feelings are insufficiently fine under the best conditions and are grossly inaccurate at times; therefore they will not do for the purpose. Thus, a polished metal surface exposed to the temperature of freezing water will feel much colder to the touch than a polished wooden surface exposed to the same conditions (for reasons to be discussed on page 225), even though both are at the same temperature. A well-known experiment produces an even greater paradox. If you place one hand in ice water and one in hot water and leave them there for a few moments, and then place both hands in the same container of lukewarm water, you will simultaneously feel the lukewarm water to be warm (with your cold hand) and cold (with your hot hand).

Some objective means of measuring temperature is therefore needed. The logical method is to find some property that changes in an apparently uniform manner with temperature and then associate fixed changes in temperature with fixed changes in that property. Physicists make use of a number of different temperature-dependent properties for the purpose, but the most commonly used property for the temperature range met with in ordinary life is that of volume-change. The volume of a given mass of matter generally increases with rising temperature and decreases with falling temperature. (I say "generally" because there are occasional exceptions to this.)

The change in volume with temperature is, in the case of liquids and solids, quite small and, in fact, unnoticeable to the eye. Thus, a steel rod a meter long will, if brought from the temperature of melting ice to that of boiling water, expand in length by one millimeter—that is, one part in a thousand. Since it is the volume that is expanding, the other dimensions will also increase by one part in a thousand, and if the steel rod has a circular cross section one centimeter in radius, that radius will increase by one-hundredth of a millimeter.

Such changes, while small, are by no means unimportant. Long metal girders such as those used in bridges, or long rails such as those used in railroad tracks, will expand and buckle in the hot summer sun if fixed at both ends. To avoid that, spaces are left between adjoining units so that there will be room to expand. Again, even a trifling change in the length of a clock's pendulum, will alter its period slightly, for that period depends upon its length (see page 112). The error in time-measurement, which depends upon that period, is cumulative and would make it necessary to adjust the clock periodically in summer, although it might run perfectly at cooler temperatures.

Not all substances expand by the same relative amounts when exposed to a given temperature change. An alloy of iron and nickel (in 5 to 3 ratio) will, for instance, only expand to one-tenth the extent that iron or steel will. For this reason, it is a useful alloy out of which to construct measuring tapes, rods of standard length, and so on. Because its length is more nearly invariable than that of most metals, the trade name for the alloy is Invar.

Glass expands with temperature almost to the same extent that steel will. If a glass vessel is exposed to a drastic temperature change, one portion of it may expand (or contract) while another portion, to which the temperature change has not yet penetrated, does not. The expansion or contraction may not be large in an absolute sense, but it is enough to set up internal strains which the cohesive forces of glass are not sufficient to withstand, so the glass cracks.

One way out of this dilemma is to use relatively thin glass so that if one portion is heated (or cooled) the temperature change will penetrate to other portions quickly.* Another and better way is to make use of a boron-containing variety of glass, usually known by the trade name of Pyrex, which will change in volume with a given temperature only by one-third the amount that ordinary glass will. It is therefore far more resistant to cracking under temperature change because smaller strains are set up. Its thickness (and mechanical strength) need not, therefore, be sacrificed to temperature stability. Quartz, with a still smaller tendency to change volume with temperature (less even than that of Invar), is still better for the purpose. A quartz vessel can be heated to red heat and plunged into ice water, and be undamaged by the ordeal.

However startling the effects of trifling changes in volume, those changes remain trifling in actual size. Unless they can somehow be magnified, they would be difficult to use as a measure of temperature. Fortunately, there are simple methods of magnifying small volume changes.

One method is to weld strips of two different metals together, say a strip of brass with one of iron. For a given change of temperature, a brass strip will change in volume (and, therefore, in

* It takes time for temperature change to make its way through glass because glass is a poor conductor (see page 224) of heat.

length) to nearly twice the extent that an iron strip of the same size will. If the two strips of metal were not welded together, the brass would expand under the influence of increasing temperature and slide against the iron, becoming a trifle the longer of the two although they had been equal in length to start with. If the temperature is reduced below the starting point, matters are reversed. Now the brass contracts more than the iron does, sliding against the iron and ending a trifle the shorter of the two.

However, the two strips of metal *are* welded together and the brass cannot slide against the iron. What happens then is that the welded strips (a *bimetallic strip* or a *compound bar*) bend in the direction of the iron if it is heated. The brass would then lie along the outer rim of the curve and the iron along the inner. Since the outer rim is longer than the inner, this allows the brass to be longer than the iron while remaining welded throughout. As the temperature falls again, the curve straightens and becomes entirely straight when the temperature returns to its original value. If the temperature falls lower still, the bimetallic strip bends in the direction of the brass, which now lies on the inner rim while the iron lies on the outer rim.

If such a bimetallic strip is fixed at one end, the other end sways back and forth as temperature changes. The outer rim of the curved strip is very little longer than the inner, and the difference in length increases but slowly with the degree of bending. For that reason, even small changes in temperature, producing very small differences in length between the iron and brass, nevertheless produce a considerable amount of bending.

A device of this sort can be used as a *thermostat*. As the temperature in the house falls, the bimetallic strip begins to curve to the left, let us say, and at a certain temperature the bending is sufficient for it to close an electrical contact that turns on the furnace. As the house heats up, the bimetallic strip bends back and quickly breaks the contact, thereby turning the furnace off. By altering the position of the electrical contact (easily done by hand), we can arrange to have the bimetallic strip turn the furnace on at any temperature we please.

Then, too, the exact position of the free end of such a bimetallic strip can be used as a measure of the temperature. If a pen is attached and a circle of paper is allowed to revolve under it at a fixed speed, the device will automatically make a continuous recording of its position, and temperatures can be deduced from the position of the wavering line.

Again, pendulums can be designed in which the rod is not

a single piece of metal but several strips of two different metals—say, steel and zinc. These can be joined by horizontal bars in such a way that the temperature change in the zinc tends to lengthen the pendulum while the temperature change in the iron tends to shorten it. The combined action tends to leave the pendulum unaltered in length as temperature changes. This is a *compensation pendulum*.

Temperature Scales

A much more common method of magnifying volume change for the purpose of measuring temperature is to make use of liquids rather than solids. Imagine an evacuated spherical container with a long, narrow tube of constant width extending upward. The container holds enough liquid to fill the sphere completely, but the neck remains empty and includes only vacuum. If the liquid is warmed, its volume will increase and there will be no place for the liquid to expand into but the neck. The volume of the water rising into the cylindrical neck can be expressed by the usual formula for the volume of a cylinder, $V = \pi r^2 h$, where r is the radius of the cylindrical neck and h is the height to which the water rises. For a given volume, the smaller the value of the radius of the neck, the greater the height to which the liquid must rise. It follows that even though the additional volume of liquid (due to expansion with temperature) is very small, the change in height can be made quite sizable if only the radius of the tube is made small enough. There is no difficulty in using changes in the height to measure changes in temperature.

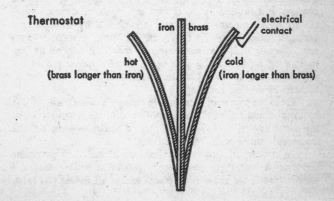

Thermostat

iron brass

electrical contact

hot
(brass longer than iron)

cold
(iron longer than brass)

A device making use of a sizable reservoir of fluid and a narrow tube into which that fluid can expand is the most common form of *thermometer* (from Greek words meaning "heat-measure"). Such thermometers were first devised in the seventeenth century and a variety of fluids were used. Water, naturally, was one of the first. Unfortunately, water does not expand uniformly with temperature. In fact, it reaches a point of maximum density and minimum volume at a temperature somewhat above its freezing point. If the temperature is dropped further, water actually expands as the temperature is lowered until it freezes (and in that process expands still further, for ice is less dense than water by nearly ten percent). Furthermore, water remains liquid over a comparatively small temperature range and is useless for temperatures below its freezing point or above its boiling point. Alcohol, also used in thermometers, stays liquid at temperatures far below that at which water freezes; however, it boils at a temperature even lower than that at which water boils.

Furthermore, both water and alcohol wet glass. As the height of liquid sinks with the falling temperature, some remains behind, clings to the glass, and then slowly trickles down. The level of liquid may then be observed to rise slowly, and it would be difficult to decide whether this was because temperature was going up or liquid was trickling down.

The first to make use of mercury as a thermometric fluid was a German physicist named Gabriel Daniel Fahrenheit (1686–1736), and he did this in 1714. Mercury's freezing point is considerably below that of water and its boiling point is considerably higher; in addition, it expands quite smoothly with temperature change. (One way of judging this is by noting that temperature changes as measured by the change in volume of mercury agree closely with temperature changes as measured by changes in a variety of other properties of matter. It is more reasonable to assume that all these changes are regular than that they all happen to be irregular in just the same way.) Finally, mercury does not wet glass, and the height of the mercury column would not be affected by delayed trickling.

Once the course of temperature change is made easily visible by the rise and fall of the mercury thread in the narrow tube of the thermometer, it is next necessary to associate some definite numerical values with fixed positions.

At atmospheric pressure, for instance, ice melts at a particular temperature and there seems every reason to believe that this temperature is the same in all places and at all times. (At least, there is no reason to believe the contrary.) Similarly, water always boils at a particular temperature at atmospheric pressure in all places and at all times. If a thermometer is placed in melting ice, the level to which the mercury thread rises can be marked; if it is placed in boiling water, a new level is marked. All men would have marks on their own thermometers that were comparable; all thermometers would match and "speak the same language." Once two such marks were set, the distance between could be divided into equal steps, or *degrees*.

Fahrenheit's method of setting his fixed point unfortunately did not involve the freezing point and boiling point of water directly. For his zero point, he used a mixture of ice and salt that produced the lowest freezing point he could get, and for another point, he tried to use the temperature of the human body. He ended by associating the freezing point of pure water with the number 32 and the boiling point of water with 212. (These figures are separated by 180 degrees, you see.) This is the *Fahrenheit scale*, and measurements upon it are given in "degrees Fahrenheit," abbreviated as "°F." Thus, the freezing point of water is 32°F, and the boiling point of water is 212°F. Body temperature is set at 98.6°F, and something like 70°F is considered a comfortable room temperature.

Temperatures below the 0°F mark can be said to be so many "degrees below zero," or a minus sign can be used. Thus alcohol freezes at 179 degrees below zero, Fahrenheit, or at −179°F.

In 1742, the Swedish astronomer Anders Celsius (1701–1744) made use of a different scale, one in which the freezing point of water was associated with 0 and the boiling point with 100, these points therefore being separated by a hundred degrees. This scale is called the *Centigrade scale* (from Latin words meaning "hundred degrees"), but in the 1950's it was decided to honor the inventor by calling it the *Celsius scale*. Whether one speaks of "degrees Centigrade" or "degrees Celsius," however, the abbreviation is "°C." On the Celsius scale, the melting point of ice, or the freezing point of water, is 0°C, and the boiling point of water is 100°C.

The Celsius scale commends itself to scientists because the 0 to 100 stretch fits in with the decimal nature of the metric system and because that stretch of temperature over which water remains liquid is a particularly interesting one, especially to chemists. It is the Celsius scale that is used universally by scientists.

The Fahrenheit scale, however, is used in ordinary affairs in the United States and Great Britain, and it has at least this advantage: the 0 to 100 stretch in the Fahrenheit scale covers the usual range of temperatures in the world. Meterologists using the Celsius scale must frequently descend to negative numbers; those using the Fahrenheit scale need do so but rarely.

Since both scales are used in the United States and Great Britain, it is useful to be able to convert one to the other. Let us begin by noting that in the range between the freezing point and the boiling point of water there are 180 Fahrenheit degrees and 100 Celsius degrees. The Celsius degree is obviously the larger of the two and is equal to 180/100, or 9/5 Fahrenheit degrees. Conversely, a Fahrenheit degree is equal to 100/180, or 5/9 Celsius degrees.

That would be enough if the two scales had their zero point in common, but they don't. If the Celsius reading is multiplied by 9/5, the result is the number of Fahrenheit degrees, not above 0°F, but above 32°F (for 32°F is the equivalent of 0°C). For that reason 32° must be added to the result. In other words:

$$F \quad \frac{9}{5}C + 32 \qquad \text{(Equation 13–1)}$$

To obtain the reading on the Celsius scale when the Fahrenheit temperature is given, it is only necessary to solve Equation 13–1 for C, and the answer is:

$$C \quad \frac{5}{9}(F - 32) \qquad \text{(Equation 13–2)}$$

Expansion

Once temperature can be measured with precision, it becomes possible to express temperature-dependent changes accurately. We can decide how much change there is "per degree Celsius" (a phrase that can be abbreviated as "per °C" or as "/°C.")

For instance, we can measure the changing length of a rod and determine the increase of length brought about by a definite temperature change. We can then calculate what the relative increase in length is for a rod that has undergone a temperature rise of 1°C. This increase is the *coefficient of linear expansion*.

The size of the coefficient of linear expansion varies from substance to substance, but for solids it is always quite small. For steel, for instance, it is 0.00001/°C, or, expressed in exponential form: 1×10^{-5}/°C. This means that a one-meter rod will expand by 0.00001 meters when temperature goes up 1°C, a one-kilometer rod will expand by 0.00001 kilometers, a one-centimeter rod will expand by 0.00001 centimeters, and so on. (Some other coefficients of linear expansion are 1.9×10^{-5}/°C for brass, 2.6×10^{-5}/°C for aluminum, and only 0.04×10^{-5}/°C for quartz.)

Suppose we represent the coefficient of linear expansion by the Greek letter "alpha" (a). If we start with a rod exactly one meter long at a particular temperature and raise that temperature by 1°C, then the length increases by a meters, and the total length is $1 + a$ meters. If we raise the temperature by 2°C, the expansion is twice as great, so the total length now becomes $1 + 2a$, while for a temperature rise of 3°C it is $1 + 3a$. In short, the value of a is multiplied by the number of degrees by which the temperature is changed.

It is customary in physics and in mathematics to signify a change in a value by the capital form of the Greek letter "delta" (Δ). If we let temperature be symbolized as t, then a temperature change is written Δt, and is usually read "delta t." In other words, we can consider the length of a one-meter rod after a certain rise in temperature to be $1 + a(\Delta t)$.

Naturally, if temperature is allowed to fall instead of rise, Δt is negative and so is $a(\Delta t)$. The expression $1 + a(\Delta t)$ is then smaller than 1, which is reasonable, since with falling temperature the rod contracts.

Suppose, now, that we started with a two-meter rod. We can consider it as consisting of 2 one-meter rods fused together. Each one-meter half has a total length of $1 + a(\Delta t)$ after the temperature has changed, and the total length is therefore $2[1 + a(\Delta t)]$. This can be reasoned similarly for any length. In fact, if we call the length of a rod L, then the new length after a change in temperature is $L[1 + a(\Delta t)]$ or, multiplying this out, $L + La(\Delta t)$.

We can next ask ourselves what the change in length is as a result of the change in temperature. The change in length, which we can naturally symbolize as ΔL, would be the length after the temperature change minus the original length. This would be $L + La(\Delta t) - L$, so we can conclude:

$$\Delta L = La(\Delta t) \tag{Equation 13-3}$$

A substance expanding with rise in temperature expands in all directions and not in length only, and the change in volume is often more important than the change in length. In liquids and gases, particularly, it is the expansion in volume that is measured. In solids, however (especially when in the shape of long rods), it is far simpler to measure the linear expansion and calculate the volume expansion from that.

We can begin by assuming that the coefficient of linear expansion for a given substance has the same value for width and height as for length.* Suppose we start with a cubic meter of a substance. Its length, after a 1°C rise in temperature becomes $1 + a$ meters. Its width, however, also expands to $1 + a$ meters, and its height, too. It's volume, which began as 1^3 cubic meters ($1^3 = 1$, of course) is now $(1 + a)^3$ cubic meters. The change in volume with a 1°C rise in temperature is $(1 + a)^3 - 1^3$, or $(1 + a)^3 - 1$, and that is the *coefficient of cubical expansion*.

The quantity $(1 + a)^3$ can be expanded by ordinary algebra to $1 + 3a + 3a^2 + a^3$. We subtract 1 from this and find that the coefficient of cubical expansion is $3a + 3a^2 + a^3$. Where a is very small, as it is in the case of solids and liquids, a^2 and a^3 are much smaller still† and can be ignored as not contributing a significant quantity to the expression. If we throw out the square and cube, then we can say with quite sufficient accuracy that the coefficient of cubical expansion is $3a$—three times the coefficient of linear expansion. Thus, if the coefficient of linear expansion is $1 \times 10^{-5}/°C$ for steel, then we can say that its coefficient of cubical expansion is $3 \times 10^{-5}/°C$.

The coefficient of cubical expansion is roughly ten times as high for liquids as for solids, and considerably higher still for gases. It is, indeed, for gases that the coefficient of cubical expansion has proved to have the greatest theoretical significance.

---

* This is not necessarily strictly true. A single crystal may expand by different amounts in different directions, depending on the orderly arrangement of the atoms and molecules making it up. A crystal may, in this respect and many others, have properties that vary with direction. In these respects, it is *anisotropic*. Common substances about us, however, are often not crystalline or, if they are, are composed of myriads of tiny crystals facing every which way. On the average, then, properties would be the same in every direction, and the substance would be *isotropic*. We tend to think of substances generally as isotropic because this is the less complicated view, but anisotropy is not really a rare phenomenon. We all know that it is much easier to split a wooden plank with the grain than against the grain.

† This may not be at once obvious. If a number is larger than 1, then the square and cube are larger still. The greater the number, the more magnified are the square and cube. Thus, the square of 10 is 100 and the cube is 1000, while the square of 100 is 10,000 and the cube is 1,000,000. The situation is reversed for numbers less than one. Here the square and cube are smaller still, and the smaller the original number the greater is the shrinkage in square and cube. Thus the square of 1/10 is 1/100 and the cube is 1/1000. For a figure like 1/100,000, which is the coefficient of linear expansion for steel, the square is 1/10,000,000,000 and the cube is 1/1,000,000,000,000,000.

Galileo himself realized that gases expand with rising temperature and contract with falling temperature; he even tried to construct a thermometer based on this fact. Taking a warmed glass bulb with an upward stalk, open at the top, he upended it in a trough of water. As the glass bulb cooled, the gas within it contracted and water was drawn part way up the stalk. Later, if the temperature went up, the gas within the bulb expanded, pushing the water level in the stalk downward. If the temperature went down, the water level rose. Unfortunately for Galileo, the water level in the stalk was also affected by changes in air pressure, so the thermometer was not an accurate one. However, the principle of change in gas volume with change in temperature was established.

Since this is so, then a volume of gas trapped under a column of mercury (as in Boyle's experiments) would expand if heated or contract if cooled. This means that if one were studying the manner in which the volume of gas changed with changes in pressure, one would have to be sure to keep the gas at constant temperature. Otherwise changes in volume would take place for which pressure was not responsible. Boyle himself in formulating what we call Boyle's law did not, apparently, take note of this fact. In 1676, however, a decade and a half after Boyle's experiments, a French physicist, Edme Mariotte (1620?-1684), discovered Boyle's law independently, and he did draw attention to the importance of constant temperature. For this reason, on the European continent the relationship of pressure and volume is often called *Mariotte's law* rather than Boyle's law—and with some justice.

The first attempt to study the expansion of gases with temperature change, quantitatively, was in 1699. The French physicist Guillaume Amontons (1663-1705) showed that if the gas were penned in and prevented from expanding as the temperature rose, the pressure increased instead, and that the pressure increased by a fixed amount for a given temperature rise regardless of the mass of gas involved.

Amontons, however, could work only with air, for in his time air was the only gas readily available. All through the eighteenth century, however, a number of gases were produced, distinguished among, and studied. In 1802, the French chemist Joseph Louis Gay-Lussac (1778-1850) not only determined the coefficient of cubical expansion for air but showed that the various common gases such as oxygen, nitrogen and hydrogen all had just about the same coefficient of cubical expansion. (This is quite astonishing, since the coefficient of cubical expansion varies quite a bit from one solid to another and from one liquid to another. Thus, the coefficient of cubical expansion is 77 times as great for aluminum as for quartz and 6 times as great for methyl alcohol as for mercury.)

The coefficient of cubical expansion for gases at 0°C turns out to be 0.00366, or about 300 times the coefficient of cubical expansion for the average solid. We can adapt Equation 13-3 for the expansion of gases. We will substitute volume (V) for length and the coefficient of cubical expansion (0.00366, or 1/273) for the coefficient of linear expansion. If we do this, then for the change in volume of gases (ΔV) with change in temperature from 0°C (Δt), we can write:

$$\Delta V = 0.00366V(\Delta t) = \frac{V(\Delta t)}{273} \qquad \text{(Equation 13-4)}$$

This is one way of expressing *Gay-Lussac's law*. As it happens, the French physicist Jacques Alexandre César Charles (1746–1823) claimed to have reached Gay-Lussac's conclusions as early as 1787. He did not publish either then or later, and ordinarily a discovery does not count unless it is published. Nevertheless, the relationship is frequently called *Charles's law* because of this.

Absolute Temperature

The fact that objects expand and contract with temperature change raises an interesting point. It is easy to see that an object can expand indefinitely as temperature goes up, but can an object contract indefinitely as temperature goes down? If it continues to contract at a steady rate, will it not eventually contract to zero volume? What then?

The paradox is most acute in the case of gases, which contract more rapidly with falling temperature than do liquids or solids. The volume of a gas after a certain change in temperature from 0°C is the original volume at 0°C plus the change in volume ($V + \Delta V$).

Suppose then that the temperature were to drop 273 degrees below 0°C. In that case, Δt would be −273. From Equation 13-4, we would see that ΔV, in that case, would be equal to $V(-273)/273$, or to $-V$. The new volume ($V + \Delta V$) would become $V - V$, or 0. A strict application of Gay-Lussac's law would indicate that gases would reach zero volume and vanish at −273°C.

Physicists did not panic at this possibility. It seemed quite likely that before −273°C was reached, all gases would be converted to liquid form, and for liquids the coefficient of cubical expansion would then be much smaller. (This turned out to be true.) Even if this were not so, it seemed quite likely that Gay-Lussac's law might not apply strictly at very low temperatures* and that the coefficient of cubical expansion might gradually decrease as temperature dropped, so although volume continued to shrink, it would do so at a slower and slower rate and never reach zero.

Nevertheless, the temperature −273°C was not forgotten. In 1848, William Thomson (later raised to the rank of baron and the title of Lord Kelvin) pointed out the convenience of supposing that −273°C might represent the lowest possible temperature, an *absolute zero*.†

If we let −273°C be zero and count upward from that by Celsius degrees, we would have an *absolute scale* of temperature. Readings on this scale would constitute an *absolute temperature*, and the degrees given in such a reading could be indicated as °A (for "absolute") or, more often, as °K (for Kelvin).

To change a Celsius temperature to one on the absolute scale, it is therefore only necessary to add 273. Since water freezes at 0°C, it does so at 273°K; since it boils at 100°C, it does so at 373°K. To prevent confusion, it is customary to represent temperature readings on the Celsius scale by the symbol t, and temperature readings on the Kelvin scale by the symbol T.** We can write the relationship of the Kelvin scale to the Celsius scale as follows, therefore:

$$T = t + 273 \qquad \text{(Equation 13–5)}$$

* It is important to remember that many scientific generalizations hold true only over limited ranges of pressure, temperature and other such environmental factors. This does not affect the usefulness of the generalization within the proper range, but one must not expect them to be useful outside that range.

† The actual value, according to the best modern determinations, is −273.16°C.

** Confusion cannot be done away with altogether. Thus, t stands not only for Celsius temperature but also, very commonly, for time. Every letter of the Latin and Greek alphabet—and some from Hebrew, Sanskrit and others—in small form, capital form, italics, boldface, and gothic script has been used, and even so there are numerous duplications of symbols. For that reason, in presenting any equation it is always advisable to state the significance of each symbol and never to take it for granted that the meaning of any symbol is self-evident.

The convenience of the absolute scale rests on the fact that certain physical relationships can be expressed more simply by using T rather than t. Thus, suppose we try to express the manner in which the volume of a quantity of gas varies with temperature. We can start at a temperature t_1, with a gas at volume V_1, and when the temperature has changed to t_2 we will find that the gas volume has changed to V_2. The final volume will be the original volume plus the volume change, so $V_2 = V_1 + \Delta V$.

Using Equation 13-4, we see that $\Delta V = V_1(\Delta t)/273$. However, the change in temperature (Δt) is the difference between the final temperature and the original temperature, $t_2 - t_1$. The unit of cubical expansion for gases is determined for a starting temperature of 0°C so $t_2 - t_1$ becomes $t_2 - 0$, or simply t_2. We will therefore substitute t_2 for the Δt in Equation 13-4. Then, in writing $V_2 = V_1 + \Delta V$, we will have:

$$V_2 = V_1 + \frac{V_1 t_2}{273} = V_1 \left(1 + \frac{t_2}{273} \right) \qquad \text{(Equation 13-6)}$$

This is easily converted to:

$$\frac{V_2}{V_1} = \frac{273 + t_2}{273} \qquad \text{(Equation 13-7)}$$

Let's consider now what the significance of the number 273 might be. It enters this equation because 1/273 is the coefficient of cubical expansion for a gas at 0°C. Remember, however, that the unit of the coefficient of cubical expansion is "per °C" or "/°C." The number 273 is the reciprocal of that coefficient, and its units should be the reciprocal of the units of the coefficient. The reciprocal of "/°C" is "°C."[*]

For 273, then, in Equation 13-7, read 273 Celsius degrees. But (see Equation 13-5) adding 273 Celsius degrees to a temperature reading on the Celsius scale gives the reading on the Kelvin scale. Consequently, the final temperature of the gas (t_2) plus 273 is the final temperature on the Kelvin scale; or, in short, $t_2 + 273 = T_2$. Similarly, 273 Celsius degrees represents the freezing point of water on the Kelvin scale, since $0 + 273 = 273$. The initial temperature of the gas was 0°C, so we can let 273 represent T_1, the initial temperature on the Kelvin scale. Consequently, Equation 13-7 becomes:

---

[*] Just as a reminder . . . The reciprocal of a is $1/a$, and the reciprocal of $1/a$ is a.

$$\frac{V_2}{V_1} = \frac{T_2}{T_1} \qquad \text{(Equation 13-8)}$$

This is another way of expressing Gay-Lussac's law (or Charles's law), and just about the simplest way. Using any other scale of temperature, the expression would become more complicated. In words, Equation 13-8 can be expressed: *The volume of a given mass of gas is directly proportional to its absolute temperature, provided the pressure on the gas is held constant.*

That last clause is important, because if the pressure on the gas varies, then the volume of the gas will change even though the temperature does not.

So we have Boyle's law which relates volume to pressure, provided temperature is held constant, and now we have Gay-Lussac's law which relates volume to temperature, provided pressure is held constant. Is there any way of relating volume to temperature *and* pressure? In other words, suppose we begin with a quantity of gas with volume V_1, pressure P_1 and temperature T_1, and change both pressure *and* temperature to P_2 and T_2. What will the new volume V_2 be?

Let's begin by changing the pressure P_1 to P_2 while holding the temperature at T_1. With the temperature constant, Boyle's law (see page 145) requires that the new volume (V_x) must fit into the following relationship: $P_2 V_x = P_1 V_1$. If we solve for V_x, we get:

$$V_x = \frac{P_1 V_1}{P_2} \qquad \text{(Equation 13-9)}$$

But V_x is not the final volume we are looking for. It is merely the volume we attain if we alter the pressure. Now let's keep the pressure at the level we have reached, P_2, and raise the temperature from T_1 to T_2. The volume now changes a second time, from V_x to V_2. (The latter is the volume we expect to have when pressure has reached P_2 and temperature has reached T_2.) In going from V_x to V_2, by raising the temperature from T_1 to T_2 and keeping the pressure constant, Gay-Lussac's law must hold, so $V_2/V_x = T_2/T_1$ (see Equation 13-8). By substituting for V_x, the value given in Equation 13-9, we have the relationship:

$$\frac{V_2}{(P_1 V_1)/P_2} = \frac{T_2}{T_1} \qquad \text{(Equation 13-10)}$$

This can be rearranged by the ordinary techniques of algebra to:

$$\frac{P_2 V_2}{T_2} = \frac{P_1 V_1}{T_1}$$

(Equation 13–11)

We can summarize then by saying that for any given quantity of gas the volume times the pressure divided by the absolute temperature remains constant. The constant here is usually symbolized as R, so we can say: $(PV)/T = R$, or:

$$PV = RT$$

(Equation 13–12)

Actual measurement, however, shows that Equation 13–12 does not hold exactly for gases (for reasons I shall explain later; see page 208). It would hold under certain ideal conditions that actual gases do not fulfill (though some come pretty close to doing so), and one can imagine an *ideal gas* or *perfect gas* that, if it existed, would follow the relationship shown in Equation 13–12 exactly. For that reason, Equation 13–12 (or its equivalent, Equation 13–11) is called the *ideal gas equation*.

CHAPTER 14

Heat

The Kinetic Theory of Gases

If the atomic theory of gas structure (made inevitable by Boyle's experiments) is to be accepted, it ought to explain the gas laws described in the previous chapter and earlier. The first man to attempt this seriously was Bernoulli (of Bernoulli's principle) in 1738.

If gases are composed of separate particles (atoms or molecules) spaced widely apart, it may reasonably be assumed that these are in constant free motion. If this were not so and the gas molecules were motionless, they would, under the force of gravity, fall to the bottom of a container and remain there. This is indeed the case for liquids and solids, where the atoms do not move freely but are in virtual contact and are constrained to remain so. The assumption that gases are made up of particles in motion, each particle virtually uninfluenced by the presence of the others, is the *kinetic theory of gases* ("kinetic," of course, from a Greek word meaning "to move").

For the moment we will not ask why the particles should be moving but will merely accept the fact that they are. The kinetic energy of the gas particles must far surpass the feeble gravitational force that can be exerted by the earth on so small a particle. (Remember that the force of gravity upon the particle depends in part upon the mass of the earth multiplied by the mass of the particle (see page 44), and the latter is so small a quantity that the total force is minute.)

To be sure, the pull of gravity is not zero and on a large scale it is effective. The earth's atmosphere remains bound to the planet by gravitational force, and most of the particles of the gas surrounding our planet remain within a few miles of the surface. Only thin wisps of gas manage to make their way higher. Nevertheless, for small quantities of gas, for quantities small enough to be contained within man-made structures, the effects of gravity are minute enough to ignore. Consequently, the particles within such containers can be viewed as moving with equal ease in any direction, upward and sideways as easily as downward.

In any given container, the random motion of the particles in any direction keeps the gas evenly spread out. (The even spreading of the gas within the container is enough to show that the motion must be random. If it were not, gas would accumulate in one part or another of the container.) If the same quantity of gas is transferred to a larger container, the random motion of the particles will spread them out evenly within the more spacious confines. Thus a gas expands to fill its container, however large, and (unless the container is so huge that the effect of gravity can no longer be ignored) fills it evenly. On the other hand, if the gaseous contents of a large container are forced into a smaller one, the particles move more closely together and all the gas can be made to fit into the smaller confine. There is none left over.

If we consider the moving gas particles, however, it is clear that no individual particle can move for long without interference. One particle is bound to collide with another sooner or later, and all are bound to collide every now and then with the walls of the container. One must assume these particles have perfect elasticity and bounce without overall loss of energy. If this were not so, the particles would gradually slow and lose energy as they bounced, until finally they were brought to a state of rest or near-rest and fell to the bottom of the container under the pull of gravity. But this does not happen. If we isolate a container of gas as best we can and keep it, that container remains full of gas indefinitely.

Bernoulli pointed out that the bouncing of gas particles off the wall of a container produced an effect that could be interpreted as pressure. As it bounced, each particle subjected the wall to a tiny force, and the total force over a unit area was the pressure. Strictly speaking, what we call pressure then is actually a great many separate pushes. There are so many of these spread so thickly through time, and each separate push is so tiny, that the whole is sensed as a smooth, even pressure. Since the particles move freely and randomly in all directions, pressure is equal in all directions.

Suppose that a gas is in a container topped by a frictionless piston with just enough weights resting upon it to balance the gas pressure (the force of the particles bouncing against the under-surface of the piston). If one of those weights is removed, the external force pressing down upon the upper surface of the piston is decreased. The upward force of the bouncing particles is greater than the downward force that remains, and the piston moves upward.

However, as the piston moves upward, the volume of the container increases. As the volume increases, each particle of the gas has, on the average, a greater distance to travel in order to reach the underside of the piston. Naturally, then, the number of collisions against the wall in any given instant must drop off as each particle spends more time traveling and less time colliding. The pressure decreases in consequence. Eventually, the pressure drops to the point where it is balanced by the fewer weights on the piston, and the piston rises no more. Gas volume has increased as pressure decreased in the manner described by Boyle's law.

Suppose, instead, that additional weights had been added to those originally present on the piston. Now the downward force of gravity moves the piston downward against the force of the particle collisions. As the piston moves downward, the volume decreases. Each particle has, on the average, a smaller distance to travel in order to reach the underside of the piston. The number of collisions in any given instant rises and pressure increases. Eventually, the pressure increases to the point where the additional weight on the piston is balanced. Gas volume has decreased and pressure has increased, again in the manner described by Boyle's law.

A century after Bernoulli's time, when the effect of temperature on the volume and pressure of gases came to be better understood, it was necessary to expand the kinetic theory of gases in order to explain the involvement of temperature.

Imagine a gas in a closed container with immovable walls. If the temperature of the gas is raised, its pressure against the walls increases. This was first observed by Amontons (see page 191) and is to be expected from the ideal gas equation (Equation

13–12), for if volume times pressure is proportional to absolute temperature, and if the volume is held constant, then pressure by itself must be proportional to the absolute temperature.

By kinetic theory, pressure increases only if the number of collisions of gas particles with the walls in any given instant increases. However, since the volume of the container (with its immovable walls) has not changed, the individual particles have the same distance to travel before reaching the walls, after the temperature rise and before. To account for the fact that more of them do reach the walls, and hence raise the pressure, one must conclude that as the temperature rises the particles move more quickly. In that case, they not only strike the wall oftener, but also more energetically. Conversely, with a fall in temperature they move more slowly.

Accepting this, let us consider a sample of gas held under a frictionless, weighted piston. The downward force of the weights is balanced by the upward force of the gas pressure. If the temperature of the gas is raised, the particles making it up move more quickly and their collisions with the underside of the piston are more numerous and energetic. The downward force of the weights is overbalanced, and the piston is raised until the expansion of volume increases the distance that must be traveled by the particles to the point where the number of collisions is so far reduced as to be only sufficient to balance the piston once more. Thus, volume increases with rising temperature. By similar reasoning, we could argue that it would decrease with falling temperature, and thus Gay-Lussac's law is explained.

I have shown how kinetic theory explains the gas laws only in a qualitative manner. In the 1860's, however, the Scottish physicist James Clerk Maxwell (1831–1879) and the Austrian physicist Ludwig Boltzmann (1844–1906) treated the kinetic theory with full mathematical rigor and established it firmly. We can consider some of this.

Let us begin with a container in the form of a parallelepiped

Kinetic theory

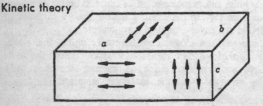

(brick-shaped, in other words), with a length equal to *a* meters, a width of *b* meters, and a height of *c* meters. The volume (V) of the container is equal to *abc* cubic meters. Suppose next that within this container are a number (N) of particles each with a mass (m), and that all the particles are moving with a velocity of *v* meters per second.

These particles can be moving in any direction, but such motion can always be viewed as being made up of three components at right angles to each other. (This can be done by setting up a "parallelepiped of force," which is a three-dimensional analog of the parallelogram of force mentioned on page 40). We can arrange the mutually perpendicular components to suit ourselves, and we can select one component parallel with the length of the container, another parallel with the width, and the third parallel with the height.

Since the motions are random and there is no net motion in any one direction (or the whole container would go flying off into space), it is fair to assume that each component contains an equal share of the motion. We suppose then that 1/3 of the total particle motion is parallel to the edge *a*, 1/3 parallel to the edge *b*, and 1/3 parallel to the edge *c*. This means that we are viewing the container of gas as containing three equal streams of particles, one moving left and right in equal amounts, one moving up and down in equal amounts, one moving back and forth in equal amounts.

In reality, of course, all the particles are continually colliding with each other, and bouncing and changing direction. Since the particles are perfectly elastic, this doesn't change the total motion, even though the distribution of motion among the individual particles is constantly changing. To put it as simply as possible, if one particle changes direction in one fashion, another particle changes simultaneously in such a way as to balance the first change. For this reason we can ignore inter-particle collisions.

Let us focus our attention on one particle moving parallel to edge *a*. It strikes the face bounded by *b* and *c* head-on and bounces back at the same speed but in the opposite direction (still parallel to edge *a*), so its velocity is now —*v*. Its momentum before the collision was *mv*, and its momentum after the collision is —*mv*. The total change in momentum is *mv* — (—*mv*), or 2*mv*.

This change in momentum must be balanced by an opposite change in momentum on the part of the wall of the container if the law of conservation of momentum is to be conserved. The wall is therefore pushed in the direction opposite to the rebounded particle, and $2mv$ represents the contribution of that one bounce of that one particle to the force on the face bounded by b and c. For the total force on the face, we need to know how many bounces there are on the entire face in a given unit of time.

The single particle we have been considering, having bounced off the face, travels to the other end of the container, bounces there, comes back, and bounces off the original face a second time; it repeats the process and bounces off a third time, then a fourth time, and so on. In traveling to the other end of the container and back, it travels a distance of $2a$ meters. Since its velocity is v meters per second, the number of its collisions with the face under discussion is $v/2a$ times each second.

The total force delivered to the wall by a single particle in one second is the momentum change in one bounce times the number of bounces per second. This is $2mv$ multiplied by $v/2a$, or mv^2/a. But one third of all the particles in the container ($N/3$) are moving parallel to edge a and each contributes the same force. The total force delivered in one second by all those particles is therefore $N/3$ multiplied by mv^2/a, or $Nmv^2/3a$.

Pressure is the force exerted against a unit area. The wall we are considering is bounded by lines of dimensions b meters and c meters, so the area of the wall is bc square meters. To get the pressure—that is, force per square meter—one must divide the total force on the wall by the number of square meters. This means we must divide $Nmv^2/3a$ by bc, and we get a pressure equal to $Nmv^2/3abc$. But abc is equal to the volume (V) of the container. We can therefore express the pressure (P) as follows:

$$P = \frac{Nmv^2}{3V} = \frac{N}{3V}(mv^2) = \frac{2N}{3V}\left(\frac{1}{2}mv^2\right) \quad \text{(Equation 14–1)}$$

But the quantity $\frac{1}{2}mv^2$ represents kinetic energy (e_k) (see page 95). We can therefore rearrange Equation 14–1 as follows:

$$PV = \frac{2N}{3}\left(\frac{1}{2}mv^2\right) = \frac{2N}{3}e_k \quad \text{(Equation 14–2)}$$

In any given quantity of gas the number of particles is constant, therefore the quantity $2N/3$ is constant. Equation 14–2 tells us, therefore, that for a given sample of gas, the product of its pressure and volume is directly proportional to the kinetic energy of its constituent particles. Equation 13–12 (see page 196) tells us, furthermore, that the product of the pressure and volume of a gas is directly proportional to its absolute temperature.

It is a truism that if x is directly proportional to y and is also directly proportional to z, then y is directly proportional to z. We conclude that if PV is directly proportional both to absolute temperature and to the kinetic energy of the gas particles, then absolute temperature is itself directly proportional to the kinetic energy of the particles of a gas (and, by extension, to the particles of any substance).

To be sure, we have assumed that all the particles in the gas have identical velocities, and that is not so. As the particles collide with each other, momentum will be transferred in a random manner (though the total momentum will always be the same). Briefly, even if the particles had originally been moving at equal velocities, they would soon be moving over a whole range of velocities.

Maxwell derived an equation that would express the distribution of particle velocities at various temperatures. If there is a distribution of velocities, there is also a distribution of kinetic energies. If we know the average velocity, however, and this can be obtained from Maxwell's equation, we know the average kinetic energy.* At any temperature, there will be individual particles with very low energies and others with very high energies. The average kinetic energy per particle, however, keeps precisely in step with the rise and fall of absolute temperature.

By the kinetic theory of gases, then, we can define heat as the internal energy associated with such phenomena as the random motions of the particles (atoms and molecules) that make up matter. Absolute temperature is the measure of the average kinetic energy of the individual particles of a system.

This gives an important theoretical meaning to absolute zero. It is not merely a convenience for simplifying equations, or a point at which the volumes of gases would shrink to zero if they followed Gay-Lussac's law exactly (which they do not). Rather it is the temperature at which the kinetic energy of the particles

---

* The average here is not the ordinary "arithmetical mean" obtained by adding values and dividing by the number of values. It is rather the "root mean square" (rms), which is the square root of the arithmetical means of the squares of the value. Thus, if we have two values, 4 and 6, the ordinary average is $(4 + 6)/2$, or 5. The rms, however is $\sqrt{\dfrac{4^2 + 6^2}{2}}$ or $\sqrt{26}$ or 5.1.

of a substance are lowered to an irreducible minimum. Usually this minimum is said to be zero, but that is not completely correct. Modern theories indicate that even at absolute zero, a very small amount of kinetic energy remains present. This amount, however, cannot be further reduced, and temperatures below absolute zero cannot exist.

Diffusion

The motion of gas particles can also be used to explain *diffusion*—that is, the spontaneous ability of two gases to mix intimately, even though originally separate, and even against the pull of gravity. Suppose a container is horizontally partitioned at the center. In the upper portion of the container is hydrogen; in the lower, under equal pressure, nitrogen. If the partition is removed, it might be expected that hydrogen (by far the lighter gas of the two) would remain floating on top, as wood floats on water. Nevertheless, in a short time the two gases are intimately mixed, the nitrogen diffusing upward, the hydrogen downward, in apparent defiance of gravity.

This comes about because the motion of the gas particles is virtually independent of gravity (see page 198). One third of the nitrogen particles (on the average, if we assume random motion) are moving upward at any one instant; and one third of the hydrogen particles are moving downward. Naturally, the two gases mix.

Diffusion also takes place between mutually soluble liquids, although more slowly. For instance, alcohol can be floated on water, which is denser; if one waits, the two liquids will mix evenly. This indicates that while the particles making up liquids must remain in contact, they nevertheless have a certain freedom of movement, slipping and sliding about, so they are able to insinuate themselves among the particles of another liquid.

On the other hand, diffusion between different solids in contact proceeds with excessive slowness, if at all, and this is an indication that the constituent particles in solids are not only in contact but are more or less fixed in place. (This does not mean, however, that the particles of a solid are motionless; all evidence points to the fact that although they have a fixed place, they vibrate about that fixed place with an average kinetic energy corresponding to the absolute temperature of the solid.)

To get a notion of the quantitative relationships that involve diffusion, let us return to Equation 14–1, where P is set equal to

$Nmv^2/3V$, and let us solve for v, the average velocity of the particles. This gives us the relationship:

$$v = \sqrt{\frac{3PV}{Nm}}$$ (Equation 14-3)

This is easily handled. If a given quantity of, let us say, oxygen is being dealt with, its pressure (P) and volume (V) are easily measured. The quantity Nm represents the number of particles multiplied by the mass of the individual particle and that is, after all, the total mass of the gas, which is easily measured. Without going into the details of the calculation, it turns out that at 0°C (273°K) and 1 atmosphere pressure,* the average velocity of the oxygen molecule is 460 meters per second (0.28 miles per second).

Equation 14-3 can be written: $v = \sqrt{3/Nm}\sqrt{PV}$. For a given quantity of a specific gas (Nm), the total mass, is constant, so the quantity $\sqrt{3/Nm}$ is constant and Equation 14-3 can be written $v = k\sqrt{PV}$, and we can say that velocity of the gas molecules is proportional to the square root of pressure times volume. By Equation 13-12 (see page 196), however, we know that PV is directly proportional to the absolute temperature T. We can therefore say that $v = k\sqrt{T}$: that the average velocity of a gas molecule is directly proportional to the square root of the absolute temperature.

If the velocity of 460 m/sec is the average for oxygen molecules at a temperature of 273°K (0°C), what would it be if the temperature were doubled to 546°K (273°C)? The average velocity is then multiplied by the $\sqrt{2}$, or approximately 1.4. Oxygen molecules, then, will move at an average velocity of 650 m/sec (0.40 miles per second) at the higher temperature.

But suppose we consider different gases at the same pressure P and volume V. Under these conditions it turns out (as a result of evidence more appropriately considered in a book on chemistry) that the number of particles present (N) is the same in both. We can consider P, V and N to be constant, therefore, and if we write Equation 14-3 as $v = \sqrt{3PV/N}\sqrt{1/m}$, we can simplify this to $v = k\sqrt{1/m}$. We are then able to say that at standard conditions

* Since many properties vary with temperature and pressure, it is usual to give the precise temperature and pressure at which a measurement is carried through. For the sake of standardization, it is common to use 0°C and 1 atmosphere pressure, or to adjust to those values if measurement is made at others. Usually, 0°C and 1 atmosphere pressure are referred to as "standard conditions of temperature and pressure," and this is abbreviated S.T.P.

of temperature and pressure the average velocity of the molecules of a gas is proportional to $\sqrt{1/m}$. Using words, we can say it is inversely proportional to the square root of the mass of individual molecules, *i.e.*, the "molecular weight."

This is sometimes called Graham's law because it was first specifically stated by the Scottish chemist Thomas Graham (1805–1869) in 1829. He noted that the rate of diffusion of a gas (which turns out to depend on the velocity of its molecules) is inversely proportional to the square root of its density (and density in gases depends on molecular weight).

A proper consideration of molecular weights is better taken up in a book on chemistry, but we can say that a molecule of hydrogen has a molecular weight 1/16 that of a molecule of oxygen. Since hydrogen molecules are 16 times less massive than oxygen molecules, they move with a velocity $\sqrt{16}$, or four times more rapidly. If at 273°K (0°C) oxygen molecules move at 460 m/sec (0.28 miles per second), then hydrogen molecules at that same temperature move at 1840 m/sec (1.12 miles per second). At 546°K (273°C) the velocity of both oxygen and hydrogen molecules is multiplied by 1.4, and the latter moves at 2600 m/sec (1.58 miles per second).

The velocity of sound through a gas depends in part upon the rapidity with which gas molecules can swing back and forth to form regions of compression and expansion (see page 165). As the molecules move more quickly with rising temperature, the velocity of sound does also. In different gases, furthermore, the velocity of sound is inversely proportional to the square root of the molecular weight, because that is the way molecular velocity varies.

Air is 4/5 nitrogen (molecular weight, 28) and 1/5 oxygen (molecular weight, 32), so the "average molecular weight" of air is 29. The molecular weight of hydrogen is 2. At a given temperature, the hydrogen molecule moves $\sqrt{29/2}$, or 3.8 times as quickly as the average molecule in air. Since at 20°C sound travels through air at the velocity of 344 m/sec, it would travel through hydrogen at that temperature at the velocity of about 1300 m/sec.

A column of air in an organ pipe, if disturbed, will vibrate at a natural frequency that depends on such things as the size of the column and the velocity of the air molecules. An organ pipe of a given size at a given temperature will therefore produce a note of a given pitch. If the pipe is filled with hydrogen, the molecules of which will move more rapidly than those of air, the same pipe at the same temperature will produce a sound of a much higher pitch. (A man who fills his lungs with hydrogen—not a recommended experiment—will find himself speaking, temporarily at least, in a shrill treble.)

Since both the pitch of the organ pipe and the velocity of sound depend on the velocity of the molecules of a gas, one can be calculated from the other. Indeed, in about 1800, the German physicist Ernst F. F. Chladni (1756–1827), sometimes called the "father of acoustics," calculated the velocity of sound in various gases (something rather difficult to measure directly) by noting the pitch of organ pipes filled with the gas (something quite easy to do).

To be sure, the actual velocities of individual molecules cover a broad range and some molecules of a particular gas move very rapidly. Even at 0°C there would be a very small fraction of the molecules in oxygen gas moving, at least temporarily, at velocities of 7 miles per second, which is some 25 times the average velocity.

It so happens that 7 miles per second is the escape velocity on the earth's surface (see page 63), and a molecule moving this quickly would be expected to leave the earth permanently. For this reason, it might seem that oxygen should constantly be "leaking" out of the atmosphere. So it is, but this should be no cause for panic. For one thing, there are very few oxygen molecules that travel at 25 times the average velocity. Of those that do, all but a vanishingly small number strike other molecules and lose the unusually high velocity, long before they can reach the upper regions of the atmosphere. Such leakage of oxygen as takes place, then, is so slow as to assure the earth its oxygen supply for billions of years to come.

In the case of hydrogen molecules, however, with four times the average velocity of oxygen molecules, a larger fraction can be expected to attain the escape velocity of 7 miles per second (since this is only about six times the average velocity of the hydrogen molecule). Here the leakage is indeed serious and the earth could not hold hydrogen in its atmosphere over the geologic eras—nor has it. There is good reason to think that the earth's atmosphere might have been rich in hydrogen to begin with, but it is all gone now.

The moon, with its much smaller escape velocity (see page 63), could not even hold any oxygen or nitrogen, if it had ever had any; in fact, it lacks an atmosphere altogether. Jupiter and the other outer planets, with larger velocities of escape, and with temperatures lower than those of the earth and the moon, can hold even hydrogen easily. The outer planets therefore have large hydrogen-filled atmospheres.

Real Gases

Boyle's law has always been accepted as useful through the three centuries during which scientists have been aware of its existence. Through the first two of those centuries, it was also (wrongly) considered exact. That Boyle's law, while useful, is only an approximation of the actual situation was first made clear by the French physicist Henri Victor Regnault (1810–1878), who in the 1850's measured the exact volumes of different gases under different pressures and found that the product of the two (PV) was not quite constant after all, even if the temperature were kept carefully constant. Under a pressure of 1000 atmospheres, the product could be twice as high as at 1 atmosphere pressure. Even when he worked with pressures that were only moderately high, he frequently found deviations of up to five percent. Furthermore, there were differences from gas to gas. Up to pressures of 100 atmospheres, hydrogen, nitrogen and oxygen deviated comparatively little from Boyle's law, while carbon dioxide deviated a good deal.

Yet Boyle's law can be derived from the kinetic theory of gases. Is the kinetic theory wrong then? No, not necessarily. However, in deriving Boyle's law from the kinetic theory of gases, it simplifies matters to make two assumptions that are not exactly true for real gases. For instance, it can be assumed that there are no attractive forces among the molecules of a gas, so the motion of one molecule can be considered completely without reference to the others. This is almost correct but not quite, for there are very weak attractive forces among the molecules of gases.

Another assumption is that the molecules are extremely small compared to the empty space separating them—so extremely small that their volume can be taken to be zero. Again, this is almost correct but not quite. The volume of the molecules is indeed very small, but it is not zero.

Now suppose we don't accept the simplifications but consider instead that when a molecule is about to strike the wall of a vessel there is a net pull backward from all the feeble intermolecular forces exerted upon the about-to-collide molecule by the other molecules. (This is a kind of gaseous surface tension, like the more familiar liquid surface tension described on page 126.) Because of this backward pull, the molecule does not strike the surface with full force and its contribution to the pressure is less than one would expect from kinetic theory if no intermolecular forces existed. To bring the pressure of the individual molecule up to the no-intermolecular-force ideal, we must add a small extra quantity of pressure (P_x). The ideal pressure (P_1) is then the actual measured pressure plus this extra quantity ($P + P_x$).

The more molecules present in the gas close to the colliding molecule (the more distant molecules contribute so little to the attractive force that they can be ignored), the greater the backward pull; the more the actual pressure (P) falls short of the ideal pressure (P_1), the greater the value P_x we must add to P in the case of this one colliding molecule. The quantity of nearby molecules is proportional to the density of the gas (D).

But pressure depends upon the total number of molecules striking the walls in a given time. The value for P_x also depends on that number. But that number in turn depends upon the density of the gas. Thus, P_x depends on the density of the gas first in connection with each per colliding molecule, then in connection with the number of colliding molecules per unit time. The total value of P_x depends upon its size per colliding molecule multiplied by the number of colliding molecules per unit time, or upon a factor proportional to density multiplied by another factor proportional to density. The total value is then proportional to the square of the density, D^2. If, on this occasion, we use a for the proportionality constant, we can say that $P_x = aD^2$.

For a given quantity of gas, density is inversely proportional to volume. The denser a gas, the less volume is taken up by a given quantity. If P_x is directly proportional to the square of the density, then, it must be inversely proportional to the square of the volume —that is, $P_x = a/V^2$. Since earlier I said that the ideal pressure was $P + P_x$, we can now write that as $P + a/V^2$.

Next, what about the matter of the finite volume of the molecules? If more and more pressure is put upon a gas, Boyle's law requires that the volume decrease steadily and get closer and closer to zero. The ideal volume (V_1) available for contraction is, if Boyle's law held perfectly, equal to all the volume (V) of the gas. But if a gas is actually put under great pressure, the molecules eventually make virtual contact. After that, there is practically no further shrinkage of volume with increase of pressure. The ideal volume available for contraction is the volume of the gas minus the volume of the molecules themselves. In other words, $V_1 = V - b$, where b represents the volume of the molecules.

The ideal gas equation (Equation 13–12, see page 196), based on the assumption of no intermolecular forces and no molecular volume, should really be expressed in terms of ideal pressure and ideal volume: $P_iV_i = RT$. If the expressions containing the actual pressure and volume are used for these ideal values, we get:

$$(P + a/V^2)(V - b) = RT \qquad \text{(Equation 14–4)}$$

This is the *van der Waals equation*, since it was first worked out by the Dutch physicist Johannes Diderik van der Waals (1837–1923) in 1873. The feeble attractive forces between gas molecules that help make this modification necessary are called *van der Waals forces*. The values for a and b in the van der Waals equation are usually quite small and differ from gas to gas, for the various gas molecules have their own characteristic volumes and exert forces of characteristic size among themselves.

The intermolecular forces in gas, while small under ordinary conditions, can be made to bring about important changes in gaseous properties. The attractive force among gas molecules increases as the molecules aproach one another, and the molecules approach more and more closely as the volume of a given quantity of gas decreases with increasing pressure. Where the attractive force is comparatively large to begin with, increased pressure can raise the force to a level higher than that which can be overcome by the kinetic energy of the gas molecules. The molecules will no longer be able to pull apart, but will cling together, and the substance will become a liquid. Gases such as sulfur dioxide, ammonia, chlorine and carbon dioxide can in this way be liquefied by pressure alone and, at this high pressure, be maintained as liquids at room temperature. (If there were no intermolecular forces, liquefaction could not take place under any circumstances. All substances would be gaseous under all conditions.)

Where the intermolecular attractive force is particularly weak, however, it is possible that even when the gas molecules are forced close enough together to touch, the attractive force will still not have increased to the point where it can keep the molecules together against the molecular motion representing their kinetic energy. For that reason, gases such as oxygen, nitrogen, hydrogen, helium, neon or carbon monoxide cannot be liquefied at room temperature under any pressure, no matter how high. During the early nineteenth century, gases of this sort therefore received the name of "permanent gases."

However, one might increase the attractive force and decrease the kinetic energy as well. If the former is brought about by increasing the pressure, the latter can be brought about by decreasing the temperature. If the temperature is brought low enough, the kinetic energy is decreased sufficiently for the attractive forces among the molecules of the so-called permanent gases to suffice to bring about liquefaction. The temperature at which such liquefaction becomes just barely possible is called the *critical temperature*. Above that critical temperature a substance can exist only as a gas. The existence of the critical temperature was first discovered by the Irish physicist Thomas Andrews (1813–1885) in 1869.

The critical temperature for oxygen is 154°K (−119°C), and it was only after oxygen was brought to a lower temperature that it became possible to liquefy it. Hydrogen, with still weaker intermolecular forces, must be lowered to a temperature of 33°K (− 240°C) before the kinetic energy of the molecules is low enough to be neutralized by those forces. The record in this respect is held by helium (not isolated on earth until 1898). Helium is the nearest approach, among real gases, to the gaseous ideal. Its critical temperature is 5°K (−268°C).

On the other hand, there are substances with intermolecular forces so great that they remain liquid at room temperature even under atmospheric pressure. (These intermolecular forces are more than mere van der Waals forces, and they will not be discussed in this book.) Water is the most common example of a substance liquid at ordinary temperatures and pressures. At a temperature of 373°K (100°C) and 1 atmosphere pressure, the intermolecular forces are overcome, thanks to the heightened kinetic energy, and water turns into its gaseous form: steam, or water vapor.* At temperatures over 100°C, water can be kept in liquid form by increasing the pressure. This means that the boiling point rises with increased pressure, a fact taken advantage of in pressure cookers. The critical temperature for water is 647°K (374°C) and it is only at temperatures above that, that liquid water cannot exist under any conditions.

Even in liquids, the attractive forces between molecules are not large enough to prevent the individual moleclules from slipping and sliding about. If, however, the temperature is lowered still further, a point is reached where the energies of the individual molecules are insufficiently large to give it even that much

* A gas that exists as such only at elevated temperatures is usually referred to as a "vapor."

freedom. The intermolecular forces are strong enough to keep the molecules firmly in place. They may vibrate back and forth, but the average position remains fixed and the substance is a solid. If the temperature of a solid is raised, the vibrations become more energetic, and at a certain temperature (depending on the size of the intermolecular forces involved) they become large enough to counter those forces to the extent of allowing the molecules to slide about; the solid has then melted, or liquefied. The melting point is only slightly affected by pressure.

The intermolecular forces of hydrogen are so weak that solid hydrogen melts at a temperature of only 14°K (-259°C), and liquid hydrogen boils (under atmospheric pressure) at a temperature of only 20°K (-253°C). Helium does better still. Its particles consist of individual atoms and the interatomic forces are so weak that even the irreducible bit of kinetic energy still present at absolute zero is enough to keep it liquid. Solid helium cannot exist at any temperature, however low, except under pressures greater than atmospheric. The boiling point of helium under a pressure of one atmosphere is 4°K (-269°C).

On the other hand, some substances possess intermolecular or interatomic forces so strong that they remain solids at ordinary temperatures and even considerably higher. The metal tungsten does not melt until a temperature of 3370°C is reached and does not boil, under atmospheric pressure, until a temperature of 5900°C is reached.

Specific Heat

So far in our discussion of heat in this chapter and the preceding one the emphasis has been on temperature, and we must avoid confusing the two. The terms "heat" and "temperature" are by no means identical. It is all too easy to assume that if one sample of water has a higher temperature than another, it is hotter and therefore has more heat. The final conclusion, however, is not necessarily true.

A thimbleful of water at 90°C is much hotter than a bathtubfull of water at 50°C, but there is more total heat in the bathtub of water. If both are allowed to stand, the thimbleful of water will have cooled to room temperature in an interval during which the bathtub-full of water would scarcely have cooled at all. The thimbleful loses its heat more quickly because, for one thing, it has far less heat, all told, to lose.

To specify, the heat content of a system is the total internal energy* of the molecules making it up, while temperature is the measure of the average translational kinetic energy of the individual molecules. In other words, heat represents a total quantity and temperature a quantity per molecule.

The difference can be made plainer, perhaps, by an analogy. Consider one liter of water poured into a tall thin cylinder so that it forms a column one meter high. Into a much broader cylinder, five liters of water are poured, and this water stands only 0.1 meters high. The water in the narrow cylinder exerts the greater pressure on the bottom of the container, but the water in the broader cylinder, exerting one-tenth the pressure, is nevertheless five times greater in volume. Volume is a total quantity, while pressure is a quantity per area. Therefore, temperature is to heat, as pressure is to volume.

It may seem that such a distinction between heat and temperature is unnecessary labor. After all, if one heats water, for instance, heat pours into it and the temperature goes up; the two rise together and why can't you use one as the measure of the other? Unfortunately, this parallel behavior of heat and temperature can be counted on only when you deal with a given quantity of a particular substance, and even then only over certain limited temperature ranges. We can see this if we compare the heat contents of two different subjects at identical temperature.

To do this, we need a unit of measurement for heat. Earlier in the book I mentioned such a unit, the calorie, in passing. Now let's go into such matters in a bit more detail.

Suppose we add heat to water, thus raising its temperature. Experiments will show that the amount of heat required to raise the temperature of water by a fixed number of degrees varies with the mass of the water receiving the heat.

We can assume, for instance, that 100 grams of boiling water contain a fixed amount of heat. If 100 grams of boiling water are poured into 5 kilograms (5000 grams) of cold water, the temperature of the cold water will rise about two Celsius degrees. If, on the other hand, the 100 grams of boiling water is poured into 10 kilograms of cold water, the temperature of the cold water will go up only one Celsius degree.

Again, the quantity of heat required to raise the temperature of a fixed mass of water varies with the number of Celsius degrees by which the temperature is raised. It takes twice as large a volume of boiling water to raise a particular quantity of cold water by 10

* The "internal energy" of a substance consists of the kinetic energy of its constituent particles plus the energy involved in the intermolecular attractions.

Celsius degrees than by 5 Celsius degrees. The unit of heat must therefore be defined in terms of a unit mass and a unit rise in temperature; as for instance, the quantity of heat required to raise the temperature of one gram of water by 1 Celsius degree. Actually, refined measurements show that the quantity of heat required to raise the temperature of one gram of water by 1 Celsius degree varies slightly according to the original temperature of the water, so the original temperature must also be included in the definition. We can say then:

One *calorie* is the quantity of heat required to raise the temperature of one gram of water from 14.5°C to 15.5°C.

We might also say that:

One thousand calories, or a *kilocalorie*, is the quantity of heat required to raise the temperature of a kilogram (1000 grams) of water from 14.5°C to 15.5°C.

Suppose now that a gram of aluminum is placed in boiling water for enough time to make certain that it has assumed the temperature of boiling water (100°C). Plunge the hot aluminum quickly into 100 grams of water at 0°C. The aluminum cools off and its heat is added to the water, raising its temperature from 0°C to about 0.22°C.

To raise the temperature of 100 grams of water by 0.22 Celsius degrees takes 100 times 0.22 or about 22 calories. The gram of aluminum, in cooling from 100°C to 0.22°C, has liberated some 22 calories. By the law of conservation of energy, we would expect that if this cooling liberated 22 calories, then adding 22 calories to the cold aluminum would bring it back up to 100°C. Roughly speaking, then, we can say that it takes 22 calories to raise the temperature of a gram of aluminum 100 Celsius degrees, and 0.22 calories to raise it 1 Celsius degree. This represents the *specific heat* of aluminum, where the specific heat of a substance is defined as the quantity of heat required to raise the temperature of 1 gram of that substance by 1 Celsius degree.

By this type of experiment one can find that the specific heat of iron is 0.11, that of copper 0.093, that of silver 0.056, and that of lead 0.03. If one calorie of heat is added to a gram of aluminum at 0°C, that amount of heat will be enough to heat it 1/0.22, or 4.5 Celsius degrees—that is, to a temperature of 4.5°C. The same amount of heat under the same conditions would raise the temperature of a gram of iron to 9°C, of copper to 11°C, of silver to 18°C, and of lead to 33°C.

Here you can see that the distinction between heat and

temperature is indeed a useful one, since the same quantity of heat may be added to a fixed mass of each of a number of differ-.ent substances and each will attain a different temperature. Temperature in itself is consequently no measure at all of total heat content. (To return to our volume-pressure analogy, this is like pouring equal volumes of water into cyindrical vessels of different diameters. The volumes may be the same; however, the final pressures will vary, and pressure is no measure at all of total volume.)

The conception of specific heat was first advanced by the Scottish chemist Joseph Black (1728–1799) in 1760.

Part of the reason for this variation of specific heat from substance to substance lies in the different masses of the atoms making up each. The lead atom is about 7.7 times the mass of the aluminum atom, the silver atom is 4 times the mass of the aluminum atom, the copper atom 2.3 times the mass of the aluminum atom, and the iron atom 2.1 times the mass of the aluminum atom.

Because of this, a given mass of lead, say 1 gram, contains only 1/7.7 times as many atoms as the same mass of aluminum. In adding heat to 1 gram of lead, you are therefore engaged in setting fewer atoms into motion and less heat is required to increase the kinetic energy of the individual atoms by enough to account for a 1 Celsius degree temperature rise. For this reason, the specific heat of lead, 0.03, is about 1/7.7 that of aluminum, 0.22. Similarly, the specific heat of silver is about 1/4 that of aluminum; the specific heat of copper is about 1/2.3 that of aluminum; and the specific heat of iron is about 1/2.1 that of aluminum.

The general rule is that for most elements the specific heat multiplied by the relative mass of its atoms yields a number that is approximately the same for all. Where the relative mass of the atoms of the different elements (the *atomic weight*) is chosen in such a way that the hydrogen atom, which is the lightest, has a weight of a trifle over 1, then the product of specific heat and atomic weight comes to about six calories for most elements.

This is known as the law of Dulong and Petit, after the French physicists Pierre Louis Dulong (1785–1838) and Alexis Thérèse Petit (1791–1820), who first advanced it in 1819.

Latent Heat

It might occur to you that temperature is very close to being a measure of heat content if only we count by atoms or molecules instead of by grams. This would be so if the law of Dulong and Petit held for all substances under all conditions, but it does not. It holds only for the solid elements and for these only in certain temperature ranges. In fact, it is possible to show cases in which heat content can change a great deal without any change in temperature at all, and that should at once put to rest any notion of using temperature as a measure of heat content.

Suppose 100 grams of liquid water at 0°C is added to 100 grams of liquid water at 100°C. After stirring, the final temperature of the mixture would be 50°C.

Next, suppose that 100 grams of ice at 0°C is added to 100 grams of liquid water at 100°C. After allowing the ice to melt and stirring the mixture (assuming that while we wait there is no over-all loss of heat to the outside world or gain of heat from it—a matter which can be arranged by insulating the system), we find that the temperature of the mixture is only 10°C.

Why should this be? Clearly the liquid water at 0°C had more heat to contribute to the final mixture than the ice at 0°C, and yet both liquid water and ice were at the same temperature. It seems reasonable to suppose that a quantity of the heat in the hot water was consumed, in the second case, in simply melting the ice; so much the less was therefore available for raising the temperature of the mixture.

Indeed, if we heat a mixture of ice and water, we find that no matter how much heat is transferred to the mixture, the temperature remains at 0°C until the last of the ice is melted. Only after all the ice is melted is heat converted into kinetic energy, and only then can the temperature of the water begin to rise. Experiment shows that 80 calories of heat must be absorbed from the outside world in order that 1 gram of ice might be melted, and that no temperature rise takes place in the process. The ice at 0°C is converted to water at 0°C.

But if the heat gained by the ice is not converted into molecular kinetic energy, what does happen to it? If the law of conservation of energy is valid, we know it cannot simply disappear.

The water molecules in ice are bound together by strong attractive forces that keep the substance a rigid solid. In order to convert the ice to liquid water (in which the molecules, as in all liquids, are free of mutual bonds to the extent of being able to slip and slide over, under, and beside each other) those forces must be countered. As the ice melts, the energy of heat is consumed in countering those intermolecular forces. The water molecules contain more energy than the ice molecules at the same temperature, not in the form of a more rapid motion or vibration, but in the form of an ability to resist the attractive forces tending to pull them rigidly together.

The law of conservation of energy requires that the energy change in freezing be the reverse of the energy change in melting. If liquid water at 0° is allowed to lose heat to the outside world, the capacity to resist the attractive forces is lost, little by little. More and more of the molecules lock rigidly into place, and the water freezes. The amount of heat lost to the outside world in this process of freezing is 80 calories for each gram of ice formed.

In short, 1 gram of ice at 0°C, absorbing 80 calories, melts to 1 gram of water at 0°C; and 1 gram of water at 0°C, giving off 80 calories, freezes to 1 gram of ice at 0°C.

The heat consumed in melting ice (or any solid, for that matter) is converted into a sort of potential energy of molecules. Just as a rock at the top of a cliff has, by virtue of its position with respect to gravitational attraction, more energy than a similar rock at the bottom of the cliff, so do freely moving molecules in liquids, by virtue of their position with respect to intermolecular attraction, possess more energy than similar molecules bound rigidly in solids.

It is the kinetic and potential energies of the molecules that, together, make up the internal energy that represents the heat content. It is the kinetic energy only that is measured by the temperature. By changing the potential energy only, as in melting or freezing, the total heat content is changed without changing the temperature.

The discoverer of the fact that heat melted ice without raising its temperature was Joseph Black, who first pointed out the significance of specific heat (see page 215). He referred to the heat consumed in melting as *latent heat*. "Latent" refers to something that is present in essence, but not in such a fashion as to be apparent or visible. This is just about synonymous with "potential," so the connection between "latent heat" and "potential energy" is clear.

Actually, the heat required to melt a gram of ice is its *latent heat of fusion* ("fusion" being synonymous with "melting"). The qualifying phrase "of fusion" is necessary, for another type of latent heat arises in connection with boiling or vaporization. In converting a gram of liquid water at 100°C to a gram of steam at 100°C, what remains of the intermolecular attractions must be completely neutralized. Only then are the molecules capable of displaying the typical properties of gases—that is, the virtually independent motion. In the earlier process of melting, only a minor portion of the intermolecular attractive force was countered, and the major portion remains to be dealt with. For this reason, the *latent heat of vaporization* of a particular substance is generally considerably higher than the latent heat of fusion for that same substance. Thus, the latent heat of vaporization of water—the amount of heat required to convert 1 gram of water at 100°C to 1 gram of steam at 100°C—is 539 calories. For water, the latent heat of vaporization is almost seven times as high as the latent heat of fusion.

The energy content of steam is thus surprisingly high. A hundred grams of water at 100°C can be made to yield 10,000 calories as it cools to the freezing point. A hundred grams of steam at 100°C, however, can be made to give up 53,900 calories merely by condensing it to water. The water produced can then give up another 10,000 calories if it is cooled to the freezing point. It is for this reason that steam engines are so useful and a "hot-water engine" would never do as a substitute. (It is also no accident that James Watt, the perfecter of the steam engine, was a student of Joseph Black.)

The latent heat of vaporization can be put to an important use. Suppose that a gas such as ammonia is placed under pressure in a closed container. If the pressure is made high enough, it will liquefy the gas (see page 210). As the ammonia liquefies, it gives up a certain amount of heat to the outside world. This heat would tend to raise the temperature of the immediate surroundings and of the ammonia itself. However, if the container of ammonia is immersed in running water, the heat evolved is carried off by that water and the liquid ammonia is no warmer than the gas had been.

If the container of ammonia is now removed from the water, and the pressure is lowered so that the liquid ammonia is free to boil again and become a gas, it must absorb an amount of heat equivalent to what it had given up before. It absorbs this heat from the nearest source—itself and its immediate neighbors. Some of the kinetic energy of its own molecules is converted into the potential energy of the gaseous state, and the temperature of the ammonia drops precipitously.

If a gas like ammonia is made part of a mechanical device that alternately compresses it and allows it to evaporate, a heat-pump will have been set up in which heat is pumped from the ammonia and anything in its near neighborhood out (by way of running water, for instance) to the world at large. If such a heat-pump is placed within an insulated box, we have a refrigerator.

The lowering of temperature with vaporization is made use of by our own bodies. The activity of the sweat glands keeps us covered with a thin film of moisture which, as it evaporates, withdraws heat from our body and keeps us cool. Water has the highest latent heat of vaporization of any common substance, so the fact that our perspiration is almost pure water means that little of it need be used, and we are ordinarily unaware of it. In hot weather the process must be accelerated, and if humid conditions cut down the rate of evaporation, perspiration will accumulate in visible quantities. We all know the feeling of discomfort that follows upon this partial breakdown of our own private refrigeration device.

CHAPTER 15

Thermodynamics

The Flow of Heat

In the previous chapter, I spoke of mixing hot and cold water and said that in the process an intermediate temperature was reached. It is easy to see that this is achieved by the physical intermingling of the molecules of the hot water (which possess a high average kinetic energy) with the molecules of the cold water (which possess a low one). The molecules of the mixture, taken as a whole, are bound to have an average kinetic energy of an intermediate value.

Gases, too, can blunt the extremes of temperature in this fashion. Warm air masses will mingle with cold air masses (and such mingling of air masses is the fount and origin of our weather), and the temperature of the earth's surface is kept at an intermediate value as a result. It might seem that the mixture of warm and cold on earth is not very efficient when one compares the frozen floes of the polar regions with the steaming jungles of the tropics. It could, however, be worse. Our moon is at the same average distance from the sun as the earth itself is, but unlike the earth it lacks an atmosphere. As a result, portions of its sunlit side grow hotter than even the earth's tropics do, and portions of its darkened side grow colder by far than an Antarctic winter.

The transfer of heat by currents of gas or liquid is known as *convection* (from Latin words meaning "to carry together").

Such actual movement of matter is not necessary for transfer of heat, however. If one end of a long metal rod is heated, the heat will eventually make itself felt at the other end of the rod. It is not to be supposed that there are currents of moving matter within the solid metal of the rod. What happens, instead, is something like this. As the end of the rod grows hot, the atoms of that portion of the metal gain kinetic energy. As long as the rod remains solid, the average position of each atom remains fixed, but each can and does vibrate about that position. As the atoms gain energy, the vibrations become more rapid, and the movements extend further from the equilibrium position. The atoms in the hottest portion of the rod, vibrating most energetically, jostle neighboring atoms, and those atoms, as a result of the impacts, vibrate more energetically themselves. In this way, kinetic energy jostles itself from atom to atom and, gradually, from one end of the rod to the other. This transfer of heat through the main body of a solid is *conduction* (from Latin words meaning "to lead together").

The fact that atoms and molecules of solids vibrate with greater amplitude as temperature rises means that each atom or molecule takes up more room. It is not surprising then that the volume of a solid, or a liquid for that matter, will increase with rising temperature and decrease with falling temperature (see page 182), even though the molecules remain in virtual contact throughout the temperature range up to the boiling point.

(This is not the only factor involved in the volume change that solids and liquids undergo with temperature. There is also the matter of the nature of the molecular arrangement. The molecular arrangement for a particular substance is usually more compact in the solid state than in the liquid state, so there is generally a sudden drop in volume—and consequent rise in density—as a substance freezes. Water is exceptional in this respect. Its molecular arrangement is less compact in the solid state than in the liquid. As a result, ice is less dense than liquid water and will float in it rather than sink to the bottom.)

Both convection and conduction are explainable in mechanical terms. In both cases, there are actual impingements of energetic atoms or molecules upon less energetic atoms or molcules, and energy is therefore transferred by direct contact. Heat can, however, be transmitted without direct contact at all. A hot object encased in a vacuum will make its heat felt at a distance, even though there is no matter surrounding it to carry this heat either by convection or by conduction. The sun is separated from us by almost 93,000,000 miles of vacuum better than any we can yet make in the laboratory, and yet its heat reaches us and is evident. Such heat seems to stream out of the hot object in all directions, like the conventional rays drawn about the sun by cartoonists. The word "ray" is "radius" in Latin, and the transference of heat across a vacuum is called *radiation*. The detailed discussion of radiation will be left for the second volume of this book.

Interest in the laws governing the movement of heat by any or all these methods grew sharp in the first part of the nineteenth century because of the growing importance of James Watt's steam engine, which depended in its workings on heat flow. In the steam engine, heat is transferred from burning fuel to water, converting the latter to steam. The heat of the steam then flows into the cold water bathing the condenser, and the steam, now minus its heat, is converted into water again. This heat flow that turned water to steam and back again somehow made available energy that could be converted into the kinetic energy of a piston, which, in turn, could be used to do work.

The study of the movement of heat (with particular attention, at first, to the workings of the steam engine) makes up that branch of physics called *thermodynamics* (from Latin words meaning "motion of heat"). Of course, all consideration of heat flow must assume, to begin with, that none of the heat will vanish into nothing or arise out of nothing. This is the law of conservation of energy, and so important is this generalization, in connection with thermodynamics in particular, that it is frequently called the *first law of thermodynamics*.

The first law of thermodynamics, however, merely states that the total energy content of a closed system is constant; it does not predict the manner in which the energy in such a system may shift from place to place. But even a little experience shows that some of the facts about such energy shifts seem to fall into a pattern.

For instance, suppose a closed system (that is, one that exchanges no energy with the outside world—giving off none and taking up none) consists of a quantity of ice placed in hot water. We can be quite certain that the ice will melt and the water will cool. The total energy has not changed; however, some of it has shifted from the hot water into the ice, and all the experience of mankind tells us that this shift is inevitable. Similarly, a red-hot stone will gradually cool, while the air in its neighborhood will gradually warm.

Such a flow of heat from a hot object to a cool object will continue until the temperature of different portions of the closed system are equal, and this is true whether heat is transferred by convection, conduction or radiation.

Faced with such facts about heat flow, the early workers in thermodynamics found matters most easily visualized if they thought of heat as a kind of fluid, and indeed this fluid even received a name—*caloric*, from a Latin word for "heat."

The flow of heat can be pictured by uses of fluid flow as an analogy. Imagine two vessels connected by a stopcock, with the water level high on the left side and low on the right. Naturally, water pressure is higher on the left than on the right, so there is a net pressure from left to right. If the stopcock is open, water will flow from left to right and continue flowing until the levels are equal on both sides. The high level will fall; the low level will rise; and the final level on both sides will be intermediate in height. Although the total water volume of the system has not changed, there has been a change in the distribution of water within the system leading to an equalization of pressure.

By changing a few key words, we can have the previous sentence read: "Although the total heat of the system has not changed, there has been a change in the distribution of heat within the system leading to an equalization of temperature." (Once again, as on page 213, we have an analogy between volume/pressure and heat/temperature.)

If we think of temperature as a kind of driving force directing the flow of heat, just as water pressure directs the flow of water, then it seems very natural, even inevitable, that heat should flow from a region of high temperature to one of low, without regard to the total heat content in each region.

Consider a gram of boiling water, for instance, and compare it with a kilogram of ice water. To freeze the kilogram of ice water, some 80,000 calories of heat must be withdrawn from it. To reduce the temperature of the gram of boiling water to the freezing point—and then freeze it—would require the withdrawal of 100 plus 80 calories; only 180 altogether. Any further cooling of the kilogram of ice obtained in the first case, as compared with the gram of ice obtained in the second, requires the withdrawal of a thousand times as much heat per Celsius degree from the former as from the latter. It is plain then that despite the difference in temperatures the total heat in the kilogram of ice water is much higher than the total heat in the gram of boiling water.

Nevertheless, if the gram of boiling water is added to the kilogram of ice water, heat flows from the boiling water into the ice water. It is not the difference in total heat content that determines the direction of heat flow. Rather, it is the difference in temperature. Again, our analogy—if in the connected vessels referred to above, the left were of narrow diameter and the right of wide diameter, water would flow from the region of smaller volume to that of greater volume. Not difference of total volume but difference of pressure would dictate the direction of water flow.

The rate at which water flowed from one portion of the system to another would depend on the size of the difference in pressure. When the stopcock is first opened, the water flows quickly, but as the difference in pressure on the two sides of the stopcock decreases, so would the rate of flow. The rate of flow becomes very small as the difference in pressure becomes small; it sinks to zero once the water "finds its level" and the differences in pressure disappear.

The flow of heat by conduction can, apparently, be pictured analogously. The rate of flow of heat from a hot region to a cold one depends in part on the difference in temperature between the two. It is conventional to calculate the quantity of heat that would flow in one second through a one-centimeter cube, where one face of the cube was 1 Celsius degree cooler than the face on the opposite side. This quantity of heat is the *coefficient of conductivity*, and it is measured in calories per centimeter per second per degree Celsius (cal/cm-sec-°C).

Even given a particular difference of water pressure, water flow might yet vary depending on whether it flowed through a wide orifice, a narrow orifice, a series of narrow orifices, a sponge, loosely-packed cotton, well-packed sand, and so on. The same is true for heat, and even where a given temperature difference is involved, heat will flow more rapidly through one substance than through another. In other words, the coefficient of conductivity varies from substance to substance.

Substances for which it is high are said to be good conductors of heat; those for which it is low are said to be poor conductors. In general, metals are good conductors of heat and nonmetals poor ones. The best conductor of heat is copper, with a coefficient of conductivity equal to 1.04 cal/cm-sec-°C. In comparison, water has a coefficient of conductivity of 0.0015 cal/cm-sec-°C, and some kinds of wood have coefficients as low as 0.00009 cal/cm-sec-°C.

It is for this reason that cold metal feels so much colder than cold wood. The metal and wood may be at equal temperatures, but heat leaves the hand much more quickly when it is in contact with the metal than with the wood. The temperature of the portion of the hand making contact with the substance drops much more rapidly in the first case. Analogously, it is safe to lift a kettle of boiling water by its wooden or plastic hand-grip, for the heat from the metal (which it is wiser not to touch) enters the wood or plastic slowly enough for loss by radiation to keep pace.

A system, completely surrounded by material of low heat conductivity, loses heat slowly to the outside world, or gains heat slowly, even though the temperature difference within and without is a great one. The system is made an island, so to speak, of a particular temperature in the midst of an outer sea of a different temperature. It is therefore insulated (from a Latin word for "island"), and a material of low heat conductivity is therefore a *heat insulator.*

Gases have low coefficients of conductivity; air, therefore, is a good heat insulator. Woolen blankets and clothes trap a layer of air in the tiny interstices between fibers; heat therefore travels from our body into the cold outer environment very slowly, and so we have a sensation of warmth that we would not otherwise have. Wool and air are not warm in themselves, but give the effect of warmness by helping us conserve our own body heat. Air alone would do equally well, if it could be relied on to remain still. The warmed air near our bodies is, however, constantly being replaced by cool air as a result of the ubiquitous air currents. Heat is carried away by convection, and a windy day feels colder than a still day at the same temperature.

All substances have coefficients of conductivity greater than zero, and there is no substance, therefore, that can qualify as a perfect insulator of heat. Suppose, though, we take the phrase "no substance" literally and surround a system with a vacuum. We would then have a better insulator than anything we could find in the realm of matter. A perfect vacuum possesses a coefficient of conductivity equal to zero, and cannot bring about heat loss through convection either. Even a vacuum is not a perfect insulator, however, for it will still serve as a pathway for the loss of heat by radiation.

Loss by radiation, however, is a slower process than loss by either conduction or convection. Consequently, some bottles are constructed with a double wall within which a vacuum is formed. Furthermore, the walls can be silvered so that any heat radiating across the vacuum, in either direction, is reflected almost entirely. In the end, passage of heat through such a vacuum flask, or "thermos bottle," is exceedingly slow. Hot coffee placed in such a flask remains hot for an extended period of time, and cold milk remains cold.

Such devices were first constructed by the Scottish chemist James Dewar (1842–1923) in 1892. He used them to store extremely frigid substances, such as liquid oxygen, under conditions that would cut down the entry of heat from outside and thus minimize evaporation. In the laboratory, these are still called "Dewar flasks" in his honor.

The Second Law of Thermodynamics

We might therefore summarize the discussion in the preceding section by saying that it is the experience of mankind that in any closed system heat will spontaneously flow from a hot region to a cold region. It seems fair to consider this *the second law of thermodynamics.*

This view of heat as a kind of fluid reached its peak in the 1820's. A rigorous mathematical analysis of heat flow according to this view was advanced in 1822 by Fourier, the devisor of harmonic analysis. This view was put to further use by another French physicist, Nicolas Léonard Sadi Carnot (1796–1832).

In 1824, Carnot analyzed the workings of a steam engine in terms that we may consider analogous to those that might be applied to a waterfall. The energy of a waterfall can be made to turn a water wheel, the motion of which can then be used to run all the devices attached to the wheel. In this way, energy of falling water is converted into work.

For a given volume of water, the amount of energy that can be converted to work depends on the distance through which the water drops—that is, upon the height of the pool of water at the bottom of the falls subtracted from the height of the cliff over which the water tumbles.

We could measure these two heights from any agreed-upon reference. Taking the level of the pool at the bottom of the falls as our standard, we could say that its height (h_1) was 0. Then, if the height of the cliff (h_2) was 10 meters higher, its height would be $+10$ meters. The distance fallen by the water would be h_2-h_1 —that is, $10 - 0$, or 10 meters.

We could also let sea level be the standard. In that case, h_1

might be $+ 1727$ meters, and h_2 would then be $+ 1737$ meters; $h_2 - h_1$ would be $1737 - 1727$, or still 10 meters. The most strictly rational zero point for height (at least on earth) would be the earth's center. In that case, the values of h_1 and h_2 might be 6,367,212 meters and 6,367,222 meters, respectively, and $h_2 - h_1$ would still be 10 meters. Indeed, we could let the top of the cliff be our zero point. If h_2 is 0, then h_1, representing the water level of the pool, ten meters lower than the cliff height, would have the value -10 meters. In that case, $h_2 - h_1$ would be $0 - (-10)$, or *still* 10 meters.

I have belabored this point in order to make it perfectly clear that it is not the absolute values of h_1 and h_2 that count in deciding the amount of work we can extract from the energy of falling water, but only the difference between them.

Furthermore, if we continue to consider the waterfall, a clear distinction can be drawn between the total energy content of the water and the available energy content. The water drops to the bottom of the waterfall and forms part of a quiet pool there. The pool by itself is not capable of turning a water wheel, yet it contains much potential energy. If a hole were dug, the water in that pool would drop further and some of its energy could be converted to work, provided that a water wheel was placed at the bottom of the hole. Ideally, a hole could be dug to the center of the earth, and then all the potential energy of the water (at least with respect to the earth) could be used. However, in actual practice no hole is dug, and only the energy of the falling water of the actual waterfall is used. That energy is available. The further potential energy of the water, counting down to the center of the earth, is present but unavailable.

We can apply this sort of reasoning to the flow of heat. In the steam engine (or in any heat engine—for example, one that might use mercury vapor instead of steam) heat flows from a hot region, the steam cylinder, to a cold region, the condenser. The heat flows from the high temperature to the low temperature, as water flows from a greater height to a lesser one. It is not the value of either the high or the low temperature which dictates the amount of energy that can be converted to work, but rather the temperature difference. It is fair, then, to represent the *available energy* in terms of the temperature difference within the heat engine. We can express this most conveniently in terms of absolute temperature (see page 193), a concept not yet fully worked out at the time of Carnot's premature death from cholera at the age of 36. If we consider the hot region of the heat engine to be at a temperature T_2 and the cold region to be at T_1, then the available energy can be represented as $T_2 - T_1$.

The cold region of the steam engine still contains heat, of course. If the condenser is at a temperature of 25°C, the water it contains (formed from the condensed steam) can, in principle, be cooled further and frozen, then cooled still further down to absolute zero; in the same way, water can be allowed to drop, in principle, to the earth's center. The *total energy* of the system would be represented by the difference between the temperature of the hot region and absolute zero—that is $T_2 - 0$, or simply T_2.

The maximum efficiency (E) of such a heat engine would be the ratio of the available energy to the total energy. If, under the conditions of the heat engine, all the energy of a system could be converted, in principle, to work, then the efficiency would be 1.0; if half the total energy could be converted into work, E would equal 0.5, and so on. Expressing available energy and total energy in terms of temperature differences, we can say then that:

$$E = \frac{T_2 - T_1}{T_2}$$ (Equation 15-1)

Thus, suppose that steam at a temperature of 150°C (423°K) is condensed to water at 50°C (323°K). The maximum efficiency would then be (423–323)/423, or 0.236. Less than a quarter of the total heat in the steam would be available for conversion into work.

What's more, even this value is reached only if the heat engine is mechanically perfect: if there are no losses of energy through friction; none through radiation of heat to the outside world, and so on. In actual practice, heat engines are considerably less efficient than the maximum predicted by Equation 15–1. What equation 15–1 does, however, is to set a maximum beyond which even mechanical perfection cannot pass.

Equation 15–1 is derived on the assumption that heat flows only from a hot region to a cold, never vice versa. It, too, is therefore an expression of the second law of thermodynamics (see page 226). The second law can therefore be viewed as setting a new kind of limitation on the utilization of energy.

The first law of thermodynamics (the law of conservation of energy) makes it plain that one cannot extract more energy from a system than the total energy present in the first place. The second law of thermodynamics maintains that it is impossible to extract more work from a system than the quantity of available energy present, and that the available energy present is invariably less than the total energy present unless a temperature of absolute zero can be attained.*

The second law of thermodynamics points out an important fact. In order to extract work from a heat engine, there must be a temperature difference. Suppose the hot region and the cold region were at the same temperature, both T_2. Equation 15–1 would then become $(T_2-T_2)/T_2$, or 0. There would be no available energy. (In the same way, no work could be done by a waterfall cascading down a height of 0 meters).

If this were not so, it would be conceivable that a ship traveling over the ocean could suck in water, make use of some of its energy content and then expel that water (cooler now than it was before) back into the ocean. All the ships in the world, and indeed all of man's other devices, could be run at the cost of a trifling fraction of the enormous quantity of energy in the ocean. The ocean would cool slightly in the process, and the atmosphere would warm, but the heat would flow back from air to water and all would be well.

If the second law of thermodynamics as expressed by Equation 15–1 is valid, however, this is impossible. To extract heat from the ocean, you would need a reservoir colder than the ocean and a refrigerating device to keep it colder than the ocean. The energy expended on refrigeration would be greater than the energy extracted from the ocean (assuming the refrigeration device to be mechanically imperfect, as it must be) and nothing would be gained. In fact, energy will have been lost. Virtually all "perpetual motion machines" worked up by hopeful inventors violate the second law of thermodynamics in one way or another. Patent offices will not even consider applications for such devices unless working models are supplied, and there seems little chance that a working model of such a device can ever be constructed.

Entropy

In the hands of Carnot, the second law of thermodynamics was of only limited application. He dealt only with heat engines and specifically omitted from consideration engines that worked by other means (by human or animal agency, for instance, or by the power of wind). Indeed, in Carnot's time, even the first law of

* It has been said that the first law of thermodynamics states, "You can't win," and that the second law of thermodynamics adds, "And you can't break even, either."

thermodynamics was not yet thoroughly understood in its broadest sense.

In the 1840's, however, when Joule had demonstrated the interconversion of heat and a variety of other kinds of energy, and Helmholtz had specifically declared the law of conservation of energy to be of universal generality (see page 100), it seemed that the second law, dictating the direction of flow of heat, might also be made universally applicable. In heat engines, a temperature difference was required before energy could be converted to work, but not all work-producing devices were heat engines. It was possible to obtain work out of some systems in which there was only one level of temperature.

Thus, work can be obtained from electric batteries where no temperature differences are involved. Here, however, there are differences in electrical potential (a matter which is not discussed in this book) that represent available energy. Again, chemical reactions can be made to do work though the final products of the reaction might be at the same temperature as the original reagents. The difference in chemical potential would represent the available energy in that case.

To make the second law of thermodynamics fully general, it must be seen to apply to electrical energy, to chemical energy, indeed to all forms of energy, and not to heat alone. In whatever form energy exists, work can only be obtained if the energy is present in a state of greater intensity in one portion of the system and lesser intensity in another portion. (In the case of heat, the intensity is measured as temperature; in other forms of energy, it is measured in other ways.) It is the difference in intensity that measures the available energy. What is left of the total energy content after the available energy is subtracted is the unavailable energy.

In 1850, the German physicist Rudolf Julius Emanuel Clausius (1822–1888) saw the true generality of Carnot's findings and announced it, specifically, as the second law of thermodynamics. (For this reason, Clausius is usually given the credit for being its discoverer.)

Now let's consider the second law again. In a heat engine, the temperature difference between the hot region and the cold region is the measure of the available energy. However, the second law states that in a closed system heat must flow from a hot region to a cold. With time, therefore, this temperature difference must decrease, for as the heat flows in the only direction it can flow, the hot region cools down and the cold region warms up. Consequently, the available energy decreases with time. Since the total energy remains constant, the unavailable energy must increase as the available energy decreases.

Of course, we might remove the restriction of a closed system so that we can allow heat to enter the hot region from outside and keep it from cooling down. We can also pump heat out of the cold region and keep it from warming up. (This is done in actual steam engines, where burning fuel keeps the steam chamber continually hot, and running cold water keeps the condenser continually cold.) It takes energy to pump heat into the hot region and out of the cold region, however. We are increasing the total energy of the system merely to keep the available energy constant. As total energy goes up while available energy remains constant, the unavailable energy goes up, too.

In short, no matter how we argue matters in the case of a heat engine, unavailable energy increases with time. We might make this increase a very slow one, if we insulate the system well enough to minimize heat flow from hot to cold. If we had a perfect insulator, we might even conceive of a situation in which the unavailable energy did not increase.

What applies to heat engines ought also apply to all work-producing devices. We might say then that the unavailable energy in any system can remain unchanged under ideal conditions, but always increases with time under actual conditions.

Clausius invented the word *entropy* (a word of uncertain derivation) to serve as a measure of the unavailability of energy. He showed that entropy could be expressed as heat divided by temperature. The units of entropy therefore are calories per degree Celsius. We can say then that the entropy of a system can remain unchanged under ideal conditions, but always increases with time under actual conditions. And this, too, is an expression of the second law of thermodynamics.

You must remember that the laws of thermodynamics apply to closed systems only. If we consider an open system, it is only too simple to find examples of apparent decreases in entropy.

In a refrigerator, for instance, heat is constantly being pumped from the cold objects within to the warm atmosphere outside in apparent defiance of the second law. A warm object, placed within the refrigerator, cools down; therefore, the available energy (represented by the temperature difference between the air outside and the object within the refrigerator) increases.

Where forms of energy other than heat are concerned, analogous "violations" of the second law of thermodynamics can be demonstrated. A man can walk uphill, increasing the available energy as measured by the difference in potential energy between himself and the bottom of the valley. Iron ore can be refined to pure iron and a spent storage battery can be charged—the former representing an "uphill movement" in chemical energy, the latter an "uphill movement" in electrical energy.

In every case cited, the system is not closed; energy is flowing into the system from outside. In order to make the second law of thermodynamics valid, the source of this outside energy must be included in the system so that it is "outside" no more.

Thus, material within the refrigerator does not spontaneously cool down (and remember that the original expression of the second law, see page 226, speaks only of a spontaneous flow of heat). Instead, the cooling takes place only because a motor is working within the refrigerator. Although the entropy of the refrigerator's interior is decreasing, that of the motor is increasing. Furthermore, the motor's increase is greater than the interior's decrease, so the net change in entropy over the entire system—the refrigerator's interior plus its motor—is an increase.

In the same way, the entropy decrease involved in converting iron ore to iron is smaller than the entropy increase involved in the burning coke and in the other reactions that bring about the refining of iron. The entropy increase in the electric generator supplying the electricity for the charging of the storage battery is greater than the entropy decrease of the storage battery itself as it is charged. The entropy decrease involved in a man walking uphill is less than the entropy increase involved in the reactions within his tissues which make the chemical energy of foodstuffs available for the effort involved in walking uphill.

This is true also of various large-scale, planet-wide processes that seem to involve a decrease in entropy. Examples of such entropy-decreasing phenomena are the uneven heating of the atmosphere, which gives rise to wind and weather; the lifting of uncounted tons of water miles high against the pull of gravity, which gives rise to rain and rivers; the conversion by green plants of carbon dioxide in the atmosphere to complicated organic compounds, which is the basis of the earth's never-ending food supply and of its coal and oil as well. It is because of these phenomena that the available energy on earth remains at approximately the same level through all its history; these phenomena also explain why we are in no danger of running out of available energy in the forseeable future.

Yet all these phenomena must not be considered in isolation, for all take place at the expense of the solar energy reaching the earth. It is solar energy that unevenly heats the atmosphere, that evaporates water, and that serves as the driving force for the photosynthetic activity of green plants. In the course of its radiation of heat and light, the sun undergoes a vast increase in entropy* —one that is much vaster than the relatively puny decreases of entropy in earth-bound phenomena.

In other words, if we include within our system all the activities that affect the system, then it turns out that the net change in entropy is *always* an increase. When we detect an entropy decrease, it is invariably the case that we are studying part of a system and not an entire one.

In actual practice we can never be sure that we are dealing with a closed system. No matter how we insulate, there are always influences from outside—energy gains and energy losses from and to the outside. All processes on the earth are affected by solar energy, and even if we consider the earth and sun together as one large system, there are gravitational and radiational influences from other planets and even other stars. Indeed, we cannot be certain that we are dealing with a truly closed system unless we take for our system nothing less than the entire universe.

In terms of the universe we can (as Clausius did) express the laws of thermodynamics with utmost generality. The first law of thermodynamics would be: *The total energy of the universe is constant.* The second law of thermodynamics would be: *The total entropy of the universe is continually increasing.*

Now suppose the universe is finite in size. It can then contain only a finite amount of energy. If the entropy of the universe (which is the measure of its unavailable energy content) is continually increasing, then eventually the unavailable energy will reach a point where it is equal to the total energy. Since the unavailable energy cannot rise beyond that point, the entropy of the universe will have reached a maximum.

In this condition of maximum entropy, no available energy remains, no processes involving energy transfer are possible, no work can be done. The universe has "run down."

---

* We might proceed to wonder how the sun was formed, for this formation must have involved a vast entropy decrease in order to make it possible for the sun to continue radiating, at the expense of a continual large entropy increase, for so many billions of years. However, to trace matters back beyond the sun would be more suitable in a book devoted to astronomy.

Disorder

Observations and experiments on heat in the first half of the nineteenth century assumed heat to be a fluid. From the very start of the century, however, evidence indicating that heat was not a fluid, but a form of motion, had begun to mount.

In 1798, for instance, Benjamin Thompson, Count Rumford (1753–1814), a Tory exile from the United States, was boring cannon in the service of the Elector of Bavaria. He noted that great quantities of heat were formed. Neither the cannon being bored nor the boring instrument used was at more than room temperature to begin with, and yet the heat developed by the act of boring was sufficient to bring water to a boil after a time; and the longer the boring was continued the more water could be boiled. It almost appeared as though the quantity of heat contained within the cannon and borer was infinite.

If heat were a fluid, and a form of matter, then to suppose it were formed in the act of boring raised a difficulty. Already, the French chemist Antoine Laurent Lavoisier (1743–1794) had established the law of conservation of matter, according to which matter could be neither created nor destroyed; and there was an increasing tendency among scientists to believe this generalization to be valid. If heat were being formed, then it must be something other than matter. To Rumford, the most straightforward possibility was that the motion of the boring instrument against the metal of the cannon was transformed into the motion of small parts of both borer and metal, and that it was this internal motion that was heat.

This notion was largely disregarded during the following decades. The assumption that small parts of an object might be moving invisibly seemed in 1800 to be just as difficult to accept as the assumption that matter was being created, perhaps even more difficult. A decade after Rumford's experimenting, however, the atomic theory was advanced and began to increase in popularity. By the internal movements of matter, one now meant the motions or vibrations of the atoms and molecules making it up, and the assumption of such motion became continually more acceptible. In the 1840's Joule's experiments in converting work to heat (see page 99) extended Rumford's observations and made the victory of the atomic motion view of heat inevitable. Finally, in the 1860's, the kinetic theory of gases and the concept of heat as a form of motion on the atomic scale were established rigorously by Maxwell and Boltzmann (see page 200).

This did not mean that the laws of thermodynamics, established in the first place on the basis of a fluid theory of heat, turned out to be false. Not at all! The laws were based on observed phenomena, and they remained valid. What had to be changed were the theories that explained why they were valid. The fluid theory of heat, to be sure, explained these phenomena very neatly,* but the atomic motion theory could be made to explain everything the fluid theory of heat could, and proved just as firm a foundation for the observation-based laws of thermodynamics.

To be sure, the view of heat as atomic motion is somewhat more difficult to picture and explain than the view of heat as a fluid. In the latter case, we can think of such familiar objects as waterfalls; in the former, the best we can do is imagine a set of perfectly elastic billiard balls bouncing about eternally in a closed chamber. One might suppose that of two theories one ought to accept the simpler, as Ockham's razor (see page 5) recommends. However, Ockham's razor is applied properly only when two or more theories explain all relevant facts with equal ease. This is not so in the present case.

If we confine ourselves to heat flow only, then it is easier to picture heat as a fluid than as atomic motion. However, if we are to explain the effect of heat on gas pressure and gas volume, if we are to explain specific heat, latent heat, and a host of other phenomena, it becomes very difficult to use the fluid theory. On the other hand, the atomic motion theory not only can explain heat flow but also all the other heat-involved phenomena.

Suppose, for instance, you have a hot body and a cold body in contact. The molecules in the hot body are, on the average, moving or vibrating more rapidly than the molecules in the cold body. To be sure, the molecules in both bodies possess a range of velocities, and there may be some molecules in the cold body that are moving more quickly than some molecules in the hot body, but this is an exceptional situation. When a molecule from the hot body (an "H molecule") collides with one from the cold body (a "C molecule") the chances are very good that it will be the H molecule that will be moving the more quickly of the two. Another way of putting it is that if a great number of H molecules

* In fact, it was just because it explained them so neatly that the fluid theory lasted as long as it did in the face of mounting evidence against it. It was distressing to have to give up something so convenient.

collide with a great number of C molecules, there will be a few cases where the C molecule is moving more rapidly than the H molecule with which it collides, but a vast preponderance of cases where it is the H molecule that is the more rapid of the two.

Now when two moving objects collide and rebound, the velocities of both may change in any of a large number of ways. These changes may be grouped into one of two classes. In the first class, the slower object may lose velocity in the process of collision while the faster object may gain velocity. The result would be that the slower object would finish by moving still more slowly, and the faster object would finish by moving still more quickly. In the second class, the slower object may gain velocity in the process of collision while the faster object may lose velocity. In the first class of collisions, the velocities become more extreme, in the second class more moderate.

There are many more ways in which a collision can belong to the second class than to the first. This means that over a large number of collisions in which velocity redistributes itself in a purely random manner, there will be many more collisions resulting in more moderate velocities than in more extreme velocities. Random collisions will bring about an "averaging out" of velocities.*

When a hot body and a cold body are in contact, a large number of H molecules collide with a large number of C molecules; the result is that after rebounding, the H molecules are moving less quickly on the whole, and the C molecules are moving more quickly. This means that the H molecules have become cooler and the C molecules warmer. There has been a flow of heat from the H molecules to the C molecules. The temperature of the portion of the hot body in contact drops, and that of the portion of the cold body in contact rises.

Such collisions continue not only at the boundary at which the hot and cold bodies meet, but also within the substance of each. In the hot body, for instance, H molecules that have been cooled

---

* This does not mean that all velocities will ultimately be exactly equal if only there are enough collisions. If two objects collide at equal velocities, it becomes very probable that there will be a gain in velocity of one at the expense of the other. Too much "averaging out" becomes very unlikely, therefore. Instead, "averaging out" proceeds only to a certain point and stops. At a particular temperature, the "averaging out" produces a range of velocities such as that predicted by the Maxwell-Boltzmann equations. A smaller and more limited range is extended to that point by collisions; a wider and more extended range is contracted to that point by collisions.

off by collisions with C molecules collide with neighboring molecules that have not been cooled off; here, too, there is a general moderation of velocities.

The result of these random collisions and random alterations of velocity throughout the entire system is that, eventually, the average velocities of the molecules in any portion of the system will be the same as in any other portion; this average will be a value that will lie between the two original extremes. (Hot and cold mix to produce lukewarm, so to speak.) Once the velocities are the same, on the average, throughout the system, collisions may continue to alter velocities, so a particular molecule may be moving quickly at one moment and slowly at another; however, the average will no longer change. The entire system having reached an intermediate equilibrium temperature, heat flow will cease.

In both the fluid theory of heat and the atomic motion theory, heat can be expected to flow spontaneously from a hot area to a cold area and this, after all, is a statement of the second law of thermodynamics. Yet there is a crucial difference between the two theories with respect to such heat flow.

In the fluid theory, the flow of heat is absolute. It is capable of going "downhill" only, and an "uphill" movement is inconceivable. In the atomic motion theory, however, the flow of heat is a statistical matter and is not absolute. The random changes of velocity as a result of random collisions will result, as a matter of extremely high probability but *not certainty*, in the flow of heat from hot to cold. It is extremely unlikely, but *not inconceivable*, that in every collision, the faster molecule may gain velocity at the expense of the slower one, so heat will flow "uphill" from cold to hot.

Maxwell tried to dramatize this possibility by visualizing a scientific fantasy. Imagine two gas-filled vessels, H and C, connected by a stopcock. The H vessel is the hotter, and its molecules move the more rapidly on the average.

But it is only on the average that H molecules move more rapidly than C molecules. Some H molecules happen to move slowly, and some C molecules happen to move rapidly. Suppose that an intelligent atom-sized creature is in control of the stopcock (this creature is usually referred to as "Maxwell's Demon"). When one of the minority of slow H molecules approaches, Maxwell's Demon opens the stopcock and lets it into the C chamber. When one of the minority of fast C molecules approaches, Maxwell's Demon opens the stopcock and lets it into the H chamber. At other times, the Demon keeps the stopcock closed. In this way, there is a slow but steady drizzle of low-velocity molecules into C and an equally slow and equally steady drizzle of high-velocity molecules into H. The average velocity of the molecules in C drops, while that in H rises—and heat flows uphill from cold to hot.

The chance of such "uphill" flows of heat (or of any other form of energy) is so fantastically small in the ordinary affairs of life that it is quite safe to ignore it. However, the shift from a condition of "certainty" to a condition of "probability" is of crucial importance. As scientists probed deeper and deeper into the subatomic world during the twentieth century, statistical analysis of events and their consequences became more and more important and the improbable (but not impossible) gains a perceptible chance of taking place, while more and more of those cause/effect combinations we usually assume to be certain have been shown to be only very, very, very probable. In short, Maxwell's statistical interpretation of heat flow marks one of the first steps in the transition from the "classical physics" of the nineteenth century (with which this volume is concerned) to the "modern physics" of the twentieth century.

And how can entropy be interpreted in the light of the atomic motion view of heat? Entropy, according to the second law of thermodynamics, always increases. Well, then, what is it that always increases as a result of molecular collisions? In a manner of speaking, moderation does. If in a system, to begin with, an accumulation of heat is concentrated in one portion and there is a deficit in another, molecular collisions increase moderation and spread the heat more evenly throughout the system. In the end, when temperature equilibrium is reached, heat is spread out as evenly as possible.

Entropy can therefore be interpreted as a measure of the evenness with which energy is distributed. This can be applied to any form of energy and not merely to heat. When an electric battery discharges, its electrical energy is more and more evenly distributed over its substance and over the material involved in the electrical flow of current. In the course of a spontaneous chemical reaction, chemical energies are more evenly distributed over the molecules involved.

What's more, the evenness of energy distribution is "most even," so to speak, when it is distributed as random motion among molecules. The conversion of any form of non-heat energy to heat represents a gain in the evenness with which energy is distributed and is, therefore, a gain in entropy.

It is for this reason that any process involving a transfer of energy is bound to produce heat as a side-product. A body in motion will produce heat as a result of friction or air resistance, and some of its kinetic energy will be spread out over the molecules with which it has come in contact. In converting electrical energy to light or to motion, heat is also produced, as we know if we touch an electric light bulb or an electric motor.

This means, in reverse, that if heat were completely converted into some form of non-heat energy, then there would automatically be a decrease in entropy. But a decrease in entropy in a closed system is so extremely unlikely that the possibility of its occurrence under ordinary conditions can be ignored. Some heat, to be sure, can be converted into other forms of energy, but only at the expense of further increasing the entropy of the remaining heat in the system. In the steam engine, for instance, the conversion of the heat energy of the steam into the kinetic energy of the pistons is a piece of decreasing entropy that is at the expense of the (still greater) increasing entropy of the burning fuel that produces the steam.

The increasing evenness with which energy is spread out can be interpreted as increasing "disorder." We interpret order as a quality characterized by a differentiation of the parts of a system: a separating of things into categories; a filing of cards in alphabetical order; a listing of things in terms of increasing quantities. To spread things out with perfect evenness is to disregard all these differentiations. A particular category of objects is evenly spread out among all the other categories, and that is maximum disorder.

For this reason, when we shuffle a neatly stacked deck of cards into random order, we can speak of an increase in entropy. And, in general, all spontaneous processes do indeed seem (in line with the second law of thermodynamics) to bring about an increase of disorder. Unless a special effort is made to reverse the order of things (increasing our own entropy), neat rooms will tend to become messed up, shining objects will tend to become dirty, things remembered will tend to become forgotten, and so on.

We thus find there is an odd and rather paradoxical symmetry to this book. We began with the Greek philosophers making the first systematic attempt to establish the generalizations underlying the order of the universe. They were sure that such an order, basically simple and comprehensible, existed. As a result of the continuing line of thought to which they gave rise, such generalizations were indeed discovered. And of these, the most powerful of all the generalizations yet discovered—the first two laws of thermodynamics—succeed in demonstrating that the order of the universe is, first and foremost, a perpetually increasing disorder.

Suggested Further Reading

Cajori, Florian, *A History of Physics*, Dover Publications, Inc., New York (1929).

Feather, Norman, *The Physics of Mass, Length and Time*, Edinburgh University Press, Edinburgh (1959).

Feynman, Richard P.; Leighton, Robert B.; and Sands, Matthew, *The Feynman Lectures on Physics* (Volume I), Addison-Wesley Publishing Co., Inc., Reading, Mass. (1963).

Miller, Franklin, Jr., *College Physics*, Harcourt, Brace & World, Inc., New York (1959).

Taylor, Lloyd W., *Physics* (Volume I), Dover Publications, Inc., New York (1941).

INDEX